BrightRED Study Guide

CfE HIGHER

CHEMISTRY

Bill Beveridge, Archie Gibb and David Hawley

BrightRED
PUBLISHING

First published in 2014 by:
Bright Red Publishing Ltd
1 Torphichen Street
Edinburgh
EH3 8HX

Reprinted with corrections 2015, 2017

A CIP record for this book is available from the British Library.

ISBN 978-1-906736-59-0

With thanks to:
PDQ Digital Media Solutions Ltd, Bungay (layout) and Anne Horscroft (editorial).
Cover design and series book design by Caleb Rutherford – e i d e t i c.

Acknowledgements
Every effort has been made to seek all copyright-holders. If any have been overlooked, then Bright Red Publishing will be delighted to make the necessary arrangements.

Permission has been sought from all relevant copyright holders and Bright Red Publishing are grateful for the use of the following:

Questions taken from Revised Higher Chemistry exam papers © Scottish Qualifications Authority (pages 75, 76, 77 & 79) (n.b. solutions do not emanate from the SQA); Avesun/iStock.com (page 80); chromatos/iStock.com (page 80); prill/iStock.com (page 80); SteveGreen1953/iStock.com (page 81); Robert West (page 81); Razvan/iStock.com (page); Simon A. Eugster (CC BY-SA 3.0)[1] (page 81); Markbob1968 (CC BY-SA 3.0)[1] (page 83); Cjp24 (CC BY-SA 3.0)[1] (page 83); Alsos/iStock.com (page 85).

(CC BY-SA 3.0)[1] http://creativecommons.org/licenses/by-sa/3.0/

An exam question taken from the Higher Biology 2010 paper, Section A, Question 16 © Scottish Qualifications Authority (n.b. solutions do not emanate from the SQA) (page 61)

Printed and bound in the UK

CONTENTS

INTRODUCING HIGHER CHEMISTRY

The Higher Chemistry course is divided into four units:

- Chemical Changes and Structure
- Nature's Chemistry
- Chemistry in Society
- Researching Chemistry

The second and third of these are full units while the first and last are half units.

THE BENEFITS OF HIGHER CHEMISTRY

Next time you visit a large supermarket, look around you. Not a product in the shop, from the paper bags used to collect organic mushrooms to the electronic goods on offer, has been produced without the hard work and help of chemists. Behind the scenes, chemists are working on all sorts of projects to keep us all supplied with the food and products we need. They could be creating fertilisers to help crops grow, formulating new types of ice-creams, developing shower gels or even designing higher performance rechargeable batteries for your mobile phone. This Higher Chemistry course equips you with the knowledge to understand what these chemists do, and even begins to equip you with the concepts and skills required to develop new products and work out how they can be manufactured for maximum commercial profit.

In the unit Chemical Changes and Structure you will learn about the fundamental ideas of bonding and intermolecular forces. These ideas allow chemists to understand and predict the properties of materials from their formulae. The unit Nature's Chemistry highlights the chemistry of key families of naturally occurring compounds. It shows you how chemists can design and manufacture novel compounds to bring the public exciting new products. The final theory unit, Chemistry in Society allows you to understand how to take an idea for a product from the lab into commercial production. By understanding equilibrium, enthalpy, percentage yield and atom economy, chemists can turn a manufacturing process from one that could be making a financial loss, to an immensely profitable operation. The final unit, Researching Chemistry is a skills-based unit. This allows you to develop the skills required to carry out chemical research and allows you to investigate the chemistry behind issues currently in the media

Whether you choose to follow a chemistry related career or not, Higher Chemistry is highly regarded by all employers as this course also develops numeracy and problem-solving skills and starts to build a good awareness of commercial considerations in manufacturing. The course will also allow you to better understand the chemistry behind the products you buy.

THE EXTERNAL ASSESSMENT

At the end of the course you will be assessed externally by two components.

Component 1: Question paper – 80% of total mark

This is involves a question paper of duration 2 hours and 30 minutes with a total allocation of 100 marks.

The exam paper is divided into two sections:
- Section 1 (Objective Test), worth 20 marks, made up of 20 multiple-choice questions.
- Section 2 (Written Paper), worth 80 marks, contains a mixture of restricted and extended response questions.

The majority of the marks will be awarded for applying knowledge and understanding. The other marks will be awarded for applying scientific inquiry, scientific analytical thinking and problem-solving skills.

contd

In addition, there will be two open-ended questions in the written paper. Each question will be awarded 3 marks and can be recognised by the phrase 'using your knowledge of chemistry'. The question will not directly assess knowledge taught during the course. Instead you are to use the knowledge you do have to suggest possible answers. There is no correct answer and marks will be awarded according to whether you have shown that you have a 'good' (3 marks), 'reasonable' (2 marks) or 'limited' (1 mark) understanding of the chemistry in the question.

Component 2: Assignment – 20% of total mark

The assignment has two stages – a research stage and a communication stage and in the course of the assignment you are required to:

- choose a relevant topic in chemistry
- state appropriate aim(s)
- research the topic by selecting relevant data/information
- carry out a risk assessment of procedure(s)
- process and present relevant data/information
- analyse data/information
- state conclusions
- evaluate your investigation
- explain the underlying chemistry of the topic researched
- present the findings of the research in a report.

The research stage will be conducted under some supervision and control and the communication stage will be conducted under a high degree of supervision.

There will be 20 marks allocated to the assignment and the majority of these will be awarded for applying scientific inquiry and analytical thinking skills. The other marks will be awarded for applying knowledge and understanding related to the topic chosen.

INTERNAL ASSESSMENT: THE UNITS

There are three content based units: Chemical Changes and Structure, Nature's Chemistry and Chemistry in Society. Your school or college will assess your knowledge of the chemistry within each of these units, usually using written tests. They will also collect evidence to show that during the course you have demonstrated the skills of scientific inquiry needed to carry out an experiment and have shown that you have appropriate problem-solving skills.

The Researching Chemistry unit is different from the other three. To pass this unit you need to show that you can investigate the chemistry behind a current news story, issue or application of chemistry. You will need to show that you can use the internet, books, videos or journals to gather information from at least two different sources. You will then need to show you can plan or design an experiment, take part in the practical work and record your results appropriately.

HOW WILL THIS GUIDE HELP YOU MEET THE CHALLENGES?

The aims of the Higher Chemistry course include developing your curiosity, interest and enthusiasm for chemistry in a range of contexts and we hope that this book helps you to meet these aims. Using this book should also help equip you with some understanding of the importance of chemistry in everyday life.

However the main aim of this book is to help you achieve success in the SQA exam by providing you with a concise coverage of the key areas of the course. Helpful hints are provided in the "Don't forget" features

The "Things to do and think about" and the online tests should also help you develop other skills you will be expected to demonstrate in the exam. These skills include applying what you have learned to new situations and being able to analyse information and solve problems.

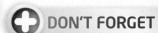 **DON'T FORGET**

A data booklet containing relevant data and formulae will be provided.

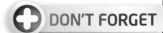 **DON'T FORGET**

The question paper will be set and marked by the Scottish Qualifications Authority (SQA).

 ONLINE

Find out more about how to research the science behind a current topic in chemistry at www.brightredbooks.net

 ONLINE

This book is supported by the BrightRED Digital Zone! Head to www.brightredbooks.net and discover a world of tests, videos, activities and more!

CONTROLLING THE RATE OF A CHEMICAL REACTION 1

Chemists must control the rates of chemical reactions. If these rates are too low, a manufacturing process will not be economically viable; too high and there is a risk of an explosion.

COLLISION THEORY

Simple **collision theory** states that, for a chemical reaction to occur:

- the reacting particles (atoms, molecules or ions) must collide into each other
- the reacting particles must collide with sufficient kinetic energy
- the reacting particles must collide with the correct alignment.

Collision theory can be used to explain the effects of concentration, pressure, surface area, temperature and collision geometry on reaction rates.

Concentration

For reactions involving solutions, increasing the concentration of a reactant speeds up the chemical reactions. At higher concentrations there are **more reacting particles in a given volume**, which means that more collisions will take place between these reacting particles and therefore the reaction is faster.

Pressure

The pressure of a gas is a measure of how often the gas molecules collide with the walls of the container. The pressure of a gas can be increased by compressing the same number of molecules into a smaller volume. For reactions involving gases, if the pressure is increased, then there will be **more reacting particles in a given volume**, which means that more collisions will take place between these reacting particles and therefore the reactions will be faster.

Surface area

In any reaction involving solids, powders react faster than larger lumps of material. If the particle size is smaller, then there is a **larger number of particles on the surface of the solid that are able to take part in collisions**. Therefore the greater the surface area, the faster the chemical reaction.

Temperature

Most chemical reactions take place faster at higher temperatures and many chemical reactions will only take place if they are heated above a certain temperature. If we consider hydrogen gas mixed with oxygen gas in a container, every second there are millions of collisions taking place between the hydrogen molecules and the oxygen molecules. However, no chemical reaction will take place until a flame or spark is used to ignite the mixture. Once started, a very fast reaction takes place.

A chemical reaction can only occur when the reacting particles collide with enough kinetic energy.

Temperature is a measure of the average kinetic energy of the particles of a substance and so **increasing the temperature means that the reacting particles will now have a greater kinetic energy and so collisions between them are more likely to be successful**.

DON'T FORGET

Decreasing the particle size of solids, increasing the concentration of solutions and increasing the pressure of gases all speed up chemical reactions because they increase the frequency of collisions between the reacting particles.

contd

Collision geometry

For a collision to result in the desired chemical reaction, the molecules must collide with the right alignment.

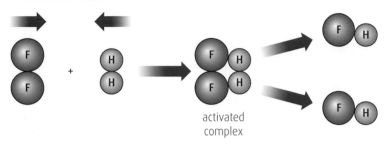

successful collision

activated complex

unsuccessful collision

MEASURING RELATIVE RATES OF REACTION

When experiments are carried out to find out the effects of changing the concentration, pressure, particle size or temperature, it is useful to have a simple way of calculating how the rate of the reaction has changed. If the time taken for the reaction to finish, or to reach a certain point, is measured, then the formula shown below can be used to calculate the 'relative rate':

$$\text{reaction rate} = \frac{1}{\text{time taken for reaction}}$$

If the time taken for the reaction is measured in seconds, then the unit for the relative rate of the reaction is s^{-1}.

THINGS TO DO AND THINK ABOUT

1   When methane gas burns in a Bunsen burner, it must first be ignited. Why does it continue to burn once it has been ignited?

2   Engineers have to take particular care when designing any part of a chemical plant in which exothermic reactions take place. If heat cannot be removed from the reaction mixture as quickly as it is generated by the reaction, then a type of explosion known as a 'thermal runaway' can occur. Why could it be dangerous to allow the temperature of an exothermic reaction mixture to increase?

 DON'T FORGET

Increasing the temperature or adding a catalyst both speed up reactions because they increase the proportion of collisions that are successful.

 VIDEO LINK

Head online and watch the clip 'Collision theory and rates of reaction' at www.brightredbooks.net

 ONLINE TEST

Take the 'Controlling the rate of a chemical reaction' test at www.brightredbooks.net

CONTROLLING THE RATE OF A CHEMICAL REACTION 2

REACTION PROFILES

A potential energy diagram is used to show the energy pathway for a chemical reaction. It shows how the potential energy changes as the reactants change into the products.

The potential energy changes for an exothermic reaction are shown diagrammatically below.

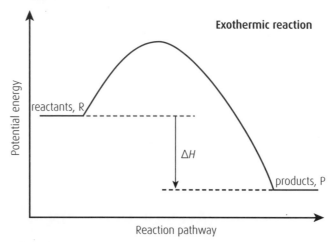

Enthalpy is a measure of the chemical potential energy contained in a substance. Enthalpy is given the symbol H. During a chemical reaction there is an enthalpy change when the reactants change into the products. It is possible to measure the enthalpy change that takes place in a chemical reaction. This is measured in kJ and when it is calculated for one mole of substance the units are kJ mol^{-1}.

The enthalpy change that takes place during a chemical reaction is given the symbol ΔH and is equal to the difference in enthalpy between the products and the reactants – that is, $\Delta H = H_P - H_R$, where H_P is the enthalpy or chemical potential energy of the products and H_R is the enthalpy of the reactants.

The enthalpy change for a chemical reaction can be calculated from the potential energy diagram for that reaction.

EXOTHERMIC AND ENDOTHERMIC REACTIONS

An **exothermic reaction** is one in which chemical potential energy is changed into heat energy. During an exothermic reaction, the temperature will be seen to rise.

As chemical potential energy is being lost as heat energy during an exothermic reaction, the value of H_P must be lower than the value of H_R.

$\Delta H = H_P - H_R$ and so, for an exothermic reaction, ΔH must have a negative value.

An **endothermic reaction** is one in which heat energy is taken in and changed into chemical potential energy. During an endothermic reaction, the temperature will be seen to drop.

As chemical potential energy is being gained from heat energy during an endothermic reaction, the value of H_P must be larger than the value of H_R.

$\Delta H = H_P - H_R$ and so, for an endothermic reaction, ΔH must have a positive value.

ACTIVATION ENERGY

In both exothermic and endothermic reactions there is a barrier which must be overcome before the reactants can change into the products. The energy difference between the chemical potential energy of the reactants and the top of this barrier is the **activation energy** for the forward reaction.

The **activation energy** is the energy required by the colliding molecules to form the activated complex.

For the collisions to be successful, the reacting particles must collide into each other with an energy equal to or greater than the activation energy. The value of the activation energy can be calculated from a potential energy diagram. In the potential energy diagram for the endothermic reaction shown below, the activation energy for the forward reaction is shown as E_A.

The activated complex is at the top of the activation barrier and is represented by ✱ in potential energy diagrams.

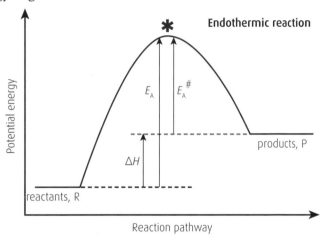

In all reactions, whether exothermic or endothermic, the activated complex always has a higher potential energy than both the reactants and the products. This is because the activated complex is an unstable arrangement of atoms compared to both the reactants and the products. The potential energy of the activated complex is always at the top of the potential energy barrier.

Consider the reversible reaction in which hydrogen and iodine react to form hydrogen iodide, and hydrogen iodide breaks down to form hydrogen and iodine. The equation for the reaction, showing both the structural formulae for the reactants and the products and the most likely activated complex, is:

$$
\begin{array}{ccccccc}
\text{H} & & \text{I} & & \text{H --- I} & & \text{H --- I} \\
| & + & | & \rightleftharpoons & \vdots \quad\; \vdots & \rightleftharpoons & + \\
\text{H} & & \text{I} & & \text{H --- I} & & \text{H --- I}
\end{array}
$$

In this equation, the solid lines represent covalent bonds and the broken lines represent covalent bonds in the process of being broken or being made.

The activated complex represents an unstable intermediate stage between the reactants and the products; it has only a fleeting existence, changing either into the products or back into the reactants.

 ## THINGS TO DO AND THINK ABOUT

State two reasons why, even if reactant molecules collide, the collision may not result in the formation of product molecules.

 **DON'T FORGET**

The same activated complex is formed in the reverse reaction as in the forward reaction.

 VIDEO LINK

Learn more about activation energy by watching the clip at www.brightredbooks.net

 ONLINE TEST

Take the 'Controlling the rate of a chemical reaction' test at www.brightredbooks.net

CONTROLLING THE RATE OF A CHEMICAL REACTION 3

KINETIC ENERGY DISTRIBUTIONS

In solids, liquids or solutions, the particles present are in continual motion. Some will be moving very slowly and others will be moving more quickly. The more quickly a particle is moving, the greater its kinetic energy.

Energy distribution diagrams

Energy distribution diagrams show how many particles are moving with each value of kinetic energy.

Point A on this graph shows that there are only a small number of slow-moving particles with low kinetic energy values.

Point D on this graph shows that there are also only a small number of very fast-moving particles with high kinetic energy values.

Point B shows that the greatest number of particles are moving with medium kinetic energy values.

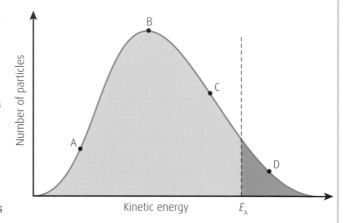

For a chemical reaction to occur, the minimum energy required by the colliding particles for a collision to be successful is known as the activation energy, E_A. Only those particles with an energy greater than or equal to the activation energy, E_A, will take part in successful collisions. For example, those particles represented at point D have an energy greater than the E_A, but those at points A, B and C do not. The total number of particles with an energy greater than the minimum activation energy required is represented by the dark green coloured area to the right of the vertical line representing the activation energy, E_A.

CHANGING THE TEMPERATURE

The effect of changing the temperature on the kinetic energy of the particles is seen in the following diagram. The curve labelled T_1 is the original curve shown above. The curve labelled $T_1 + 10°C$ shows the energy distribution when the temperature has increased by 10°C.

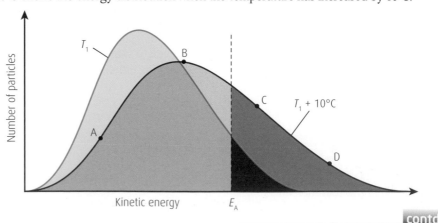

contd

DON'T FORGET

On an energy distribution diagram, the area under the curve to the right of E_A represents the number of particles which have sufficient energy to react.

VIDEO LINK

Head to the Digital Zone to watch video 'Kinetics: Chemistry's demolition derby' at www.brightredbooks.net

DON'T FORGET

Increasing the temperature means that more particles have an energy greater than the activation energy.

At the higher temperature of $T_1 + 10°C$, many more particles now have an energy equal to or greater than the activation energy for the reaction. For example, the particles at point C now have enough energy for successful collisions. The relatively small increase in temperature has caused the coloured area under the curve to the right of E_A, which represents the number of molecules with sufficient energy to react, to increase significantly.

USING A CATALYST

A catalyst speeds up a chemical reaction. The catalyst takes part in the reaction, but is regenerated at the end of the reaction. In other words, the catalyst is the same at the end as it was at the beginning of the reaction.

All catalysts provide an alternative reaction pathway with a lower activation energy. Often a difficult single-step reaction is replaced by a series of much easier reactions. Less energy is needed and so the activation energy for the reaction is lowered.

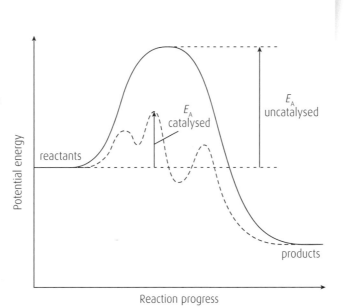

The effect that this has on the number of reacting particles which can have successful collisions is shown below. The broken red line shows the activation energy for the reaction when no catalyst is present and only those particles in the area coloured orange have enough energy for reaction.

The broken purple line represents the lowered activation energy when a catalyst is used. Now the particles within the area coloured pink, as well as those within the area coloured orange, have an energy greater than the activation energy for the catalysed reaction. Now there are more successful collisions and therefore a faster rate of reaction.

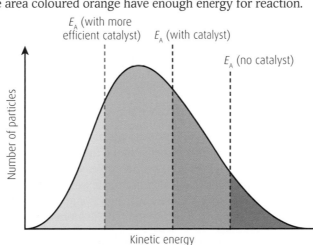

The broken blue line represents the lowered activation energy when an even more efficient catalyst is used. Now the particles within the area coloured blue, as well as those within the areas coloured orange and pink, have an energy greater than the activation energy for the reaction.

 THINGS TO DO AND THINK ABOUT

Explain the different ways in which a catalyst increases the number of reacting particles with the minimum energy required (the activation energy) compared with how increasing the temperature increases the number of particles with the minimum energy required.

 DON'T FORGET

Catalysts provide an alternative reaction pathway with a lower activation energy.

 VIDEO LINK

Watch the video about catalysis at www.brightredbooks.net

ONLINE TEST

Take the 'Controlling the rate of a chemical reaction' test at www.brightredbooks.net

PERIODICITY 1

ELEMENTS OF THE PERIODIC TABLE

The modern Periodic Table lists the elements arranged in order of increasing atomic number.

The vertical columns are known as groups and the horizontal rows are known as periods.

1 H hydrogen																	2 He helium
3 Li lithium	4 Be beryllium			symbol / name								5 B boron	6 C carbon	7 N nitrogen	8 O oxygen	9 F fluorine	10 Ne neon
11 Na sodium	12 Mg magnesium											13 Al aluminium	14 Si silicon	15 P phosphorus	16 S sulfur	17 Cl chlorine	18 Ar argon
19 K potassium	20 Ca calcium	21 Sc scandium	22 Ti titanium	23 V vanadium	24 Cr chromium	25 Mn manganese	26 Fe iron	27 Co cobalt	28 Ni nickel	29 Cu copper	30 Zn zinc	31 Ga gallium	32 Ge germanium	33 As arsenic	34 Se selenium	35 Br bromine	36 Kr krypton
37 Rb rubidium	38 Sr strontium	39 Y yttrium	40 Zr zirconium	41 Nb niobium	42 Mo molybdenum	43 Tc technetium	44 Ru ruthenium	45 Rh rhodium	46 Pd palladium	47 Ag silver	48 Cd cadmium	49 In indium	50 Sn tin	51 Sb antimony	52 Te tellurium	53 I iodine	54 Xe xenon
55 Cs caesium	56 Ba barium	71 Lu lutetium	72 Hf hafnium	73 Ta tantalum	74 W tungsten	75 Re rhenium	76 Os osmium	77 Ir iridium	78 Pt platinum	79 Au gold	80 Hg mercury	81 Tl thallium	82 Pb lead	83 Bi bismuth	84 Po polonium	85 At astatine	86 Rn radon
87 Fr francium	88 Ra radium	103 Lr lawrencium	104 Rf rutherfordium	105 Db dubnium													

metals this side ← → non-metals this side

KEY: yellow = alkali metals, orange = halogens, green = noble gases.

All the elements in any one group have similar chemical properties because their atoms have the same number of electrons in the outer shell. For example, all the elements in Group 1, the alkali metals, react vigorously with water to produce hydrogen and an alkaline solution. The elements in Group 7, the halogens, are very reactive non-metals which react with metals to form salts. The elements in Group 0, the noble gases, are very unreactive and only rarely form compounds.

ONLINE

Head online and check out the 'Royal Society of Chemistry's Periodic Table' link for a great interactive resource at www.brightredbooks.net

MONATOMIC ELEMENTS

The noble gases, Group 0, exist as single atoms (monatomic). The noble gases all have a very stable outer electron shell. Among the first 20 elements, those which exist as monatomic gases are: helium, He; neon, Ne; and argon, Ar. Because these elements do not form normal chemical bonds, scientists were surprised to discover that, at extremely low temperatures, they can turn into liquids and solids. At these very low temperatures there must be some force acting to hold the atoms together.

These weak forces are called **London dispersion forces**. London dispersion forces can operate between all atoms and molecules and are caused by the movement of electrons inside atoms and molecules. At any one instant the electron distribution in an atom or molecule might be such that there is a very slightly negative charge at one side of the atom or molecule. There will be a corresponding very slightly positive charge at the other side. This unequal distribution of charge means that a temporary dipole has been created. This is known as an **instantaneous dipole**.

ONLINE

Learn more about electron density online at www.brightredbooks.net

If one end of an atom or molecule is slightly negative, then it will repel the negative electrons on adjacent atoms or molecules, causing dipoles there. Temporary dipoles formed this way are said to be **induced dipoles**.

The positive end of a dipole on one atom or molecule is attracted to the negative end of a dipole on a nearby atom or molecule.

contd

London dispersion forces are a result of the electrostatic attractions between instantaneous dipoles and induced dipoles caused by the movement of electrons inside atoms and molecules.

The strength of the London dispersion forces is proportional to the number of electrons in an atom or molecule. The more electrons that are present, the stronger the London dispersion forces are between atoms. These London dispersion forces have to be overcome before the monatomic elements will melt or boil. This explains why there is a definite increase in the boiling points of the elements on moving down Group 0, as shown in the graph below.

There is no bonding involved in the noble gases except for the very weak London dispersion forces between the atoms when they are close together in the liquid and solid states. These forces are easily overcome and so the noble gases have very low melting and boiling points and are all gases at room temperature.

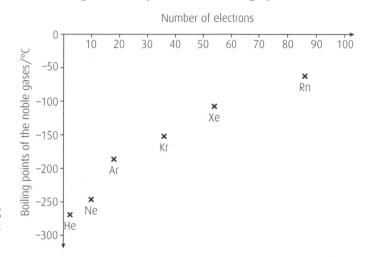

METALLIC ELEMENTS

Metallic elements are found on the left hand side of the Periodic Table. Metal atoms readily lose their outer electrons. The structure of metals in the solid state is therefore regarded as a regular array or lattice of positive ions held together by a 'sea' of delocalised electrons.

A metallic bond consists of the electrostatic attraction between the positive metal ions and the negative delocalised electrons. This holds the metallic structure together as a single unit. The delocalised outer electrons are free to move throughout the metallic structure or lattice. The presence of these delocalised electrons explains why metals are very good conductors of heat and electricity. Other properties of metals explained by the delocalised electrons are:

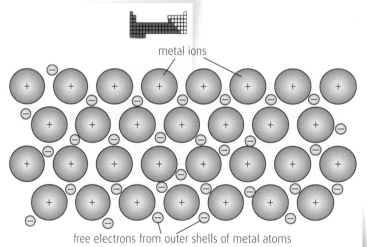

metal ions

free electrons from outer shells of metal atoms

- ductility – can be drawn into wires
- malleability – can be beaten into sheets.

Metallic bonding is fairly strong and so metals generally have high melting and boiling points. An exception is mercury. The strength of the metallic bond depends to some extent on the number of outer electrons. The alkali metals (Group 1) have only one outer electron and have lower melting and boiling points than most other metals.

THINGS TO DO AND THINK ABOUT

There are only seven naturally occurring noble gases, but scientists working at the Joint Institute for Nuclear Research in Dubna, Russia, have reported that they have been able to make a few atoms of a new element, recently named as oganesson. This new noble gas has the atomic number 118. Using the pattern in boiling points shown in the above graph, predict the boiling point for oganesson.

PERIODICITY 2

COVALENT ELEMENTS

Most non-metallic elements can form covalent bonds. Covalent bonds are formed by the merging or overlapping of half-filled outer electron clouds between the positive nuclei of two atoms. The positive nuclei of both atoms attract the electrons in the overlap region and this is what holds the two atoms together.

COVALENT MOLECULAR ELEMENTS

Covalent molecular elements are made up of molecules. A **molecule** is a group of atoms held together by covalent bonds.

Elements 1–20 with covalent molecular structures include:

- hydrogen, H−H; nitrogen, N≡N; oxygen, O=O; fluorine, F−F; and chlorine, Cl−Cl
- phosphorus, with P_4 molecules
- sulfur, with S_8 molecules
- carbon, in its fullerene form, with molecules such as C_{60}.

Elements which have covalent molecular structures have low melting and boiling points because there are only very weak London dispersion forces between the molecules. Some of these must be overcome at the melting point when the solid becomes a liquid and all of them have to be overcome at the boiling point when the liquid changes to a gas. The strong covalent bonds inside the molecules are not broken at the melting or boiling points of the elements.

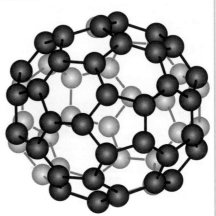

Fullerenes are molecular forms of carbon. This one, with 60 atoms, has 360 electrons in each molecule, so the boiling point is relatively high.

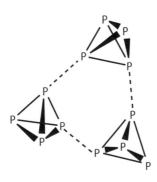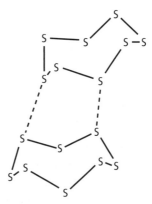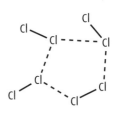

P_4 molecules
Four atoms × 15 e⁻ per atom
60 electrons in each molecule
Boiling point 280°C

S_8 molecules
Eight atoms × 16 e⁻ per atom
128 electrons in each molecule
Boiling point 445°C

Cl_2 molecules
Two atoms × 17 e⁻ per atom
34 electrons in each molecule
Boiling point −35°C

COVALENT NETWORK ELEMENTS

A covalent network structure consists of a giant lattice of covalently bonded atoms. There are three elements in the first 20 elements of the Periodic Table which have covalent network structures. These are boron, silicon and carbon (as diamond or graphite). These elements have very high melting and boiling points because the covalent bonds, which are very strong, have to be broken at their melting and boiling points. They are very hard materials because it is not possible to move any of the atoms in the structure without breaking the very strong covalent bonds.

The diagram on the right shows part of the structure of graphite, another pure form of carbon. The carbon atoms form layers of hexagonal rings. Carbon has four electrons in its outer shell. Three of these electrons are used to form covalent bonds. The spare electron in each carbon atom becomes delocalised within the layer, just like the delocalisation of the outer electrons in metallic structures. This explains why graphite is a good conductor of electricity.

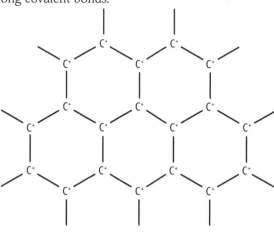

The layers of carbon atoms in graphite are attracted to the layers above and below by London dispersion forces as shown in this diagram. These are much weaker than the strong covalent bonds so, although the atoms within each layer are held together strongly, the layers can easily slide across each other. Graphite is used as the 'lead' in pencils because, as it moves across the paper, layers rub off leaving a mark on the paper. Graphite is also used as a lubricant.

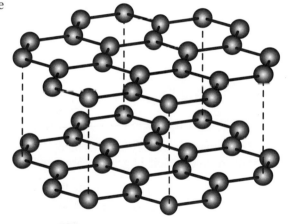

Silicon and diamond (a pure form of carbon) have similar structures to each other, with each atom covalently bonded to another four atoms in a gigantic tetrahedral arrangement of atoms as shown in the diagram on the right.

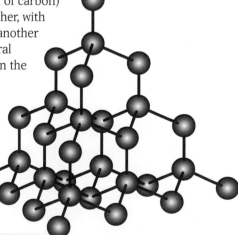

Boron is another non-metal with a covalent network structure.

DON'T FORGET

Elements with covalent network structures have very high melting points because every atom is locked in place by strong covalent bonds.

 VIDEO LINK

For a great video about carbon, head to www.brightredbooks.net

THINGS TO DO AND THINK ABOUT

Which bonds or intermolecular forces are being broken when the following elements melt: sulfur, zinc, phosphorus, argon, chlorine, carbon in the form of diamond, carbon in the form of a fullerene?

 ONLINE TEST

Test yourself on periodicity online at www.brightredbooks.net

PERIODICITY 3

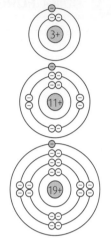

Arrangement of electrons in elements in Group 1.

ATOMIC SIZE

The **covalent radius** of an element is half the distance between the nuclei of two of its bonded atoms. The covalent radius of each element given in the Data Booklet (page 7) is taken to be a measure of the size of the atoms of that element.

The **covalent radius increases down a group**. This is because, on moving down a group from one element to the next, the number of occupied electron shells, or energy levels, increases.

The covalent radius across a period decreases from left to right. The atoms of all the elements in a period (horizontal row) have the same number of electron shells, but the nuclear charge is increasing. Moving across a period from left to right there is an increase of one proton in the nucleus from one element to the next. This increase in nuclear charge exerts an increasing attraction on the outer electrons, pulling them in closer to the nucleus and decreasing the covalent radius.

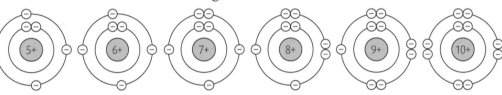

Arrangement of electrons in elements in Period 2.

IONISATION ENERGY

The ionisation energy measures how tightly an atom holds on to its outer electrons.

The **first ionisation energy** of an element is the energy required to remove one mole of electrons from one mole of gaseous atoms of the element.

DON'T FORGET

The equation must show that both the atom and the ion formed are in the gaseous state.

The general equation representing the first ionisation energy is given in the Data Booklet (page 11) and is $E(g) \rightarrow E^+(g) + e^-$, in which E represents any element.

This means that the equation for the first ionisation energy for sodium is $Na(g) \rightarrow Na^+(g) + e^-$ and, for chlorine, $Cl(g) \rightarrow Cl^+(g) + e^-$.

The second and subsequent ionisation energies refer to the energies required to remove further moles of electrons.

For example, the second ionisation energy refers to the equation $E^+(g) \rightarrow E^{2+}(g) + e^-$, so the equation representing the second ionisation energy for potassium is $K^+(g) \rightarrow K^{2+}(g) + e^-$.

The third ionisation energy refers to the equation $E^{2+}(g) \rightarrow E^{3+}(g) + e^-$, so the equation representing the third ionisation energy of aluminium is $Al^{2+}(g) \rightarrow Al^{3+}(g) + e^-$.

Notice that in each equation representing an ionisation, the reactant and the product are both in the gaseous state and that only one mole of electrons is removed each time.

Trends in ionisation energies

Ionisation energy values are given in the Data Booklet (page 11).

The ionisation energy increases across a period from left to right – that is, more energy is required to remove an electron on moving from left to right. The reasons for this include:

● the positive **nuclear charge is increasing**, so it is more difficult for the negative electron to be 'pulled' away

contd

- the atoms are getting smaller, so the negative electron is closer to the positive nucleus and is attracted more strongly.

The ionisation energy decreases down a group – that is, less energy is required to remove an outer electron on moving down a group. The reasons for this include:

- the outer electron to be removed is in a shell **further away** from the nucleus, so the atoms are getting bigger and the negative electron is further away from the positive nucleus and is therefore attracted less strongly

- there will a **greater screening effect** as there are more shells of inner electrons between the outer shell and the nucleus – these inner electrons are said to 'screen' or 'shield' the electrons in the outer shell from the positive nucleus.

ONLINE

Download and study the slideshow 'Trends in the Periodic Table and Bonding' at www.brightredbooks.net

ELECTRONEGATIVITY

A covalent bond occurs when a pair of electrons is shared between two atoms. If the two atoms are different sizes, or have different charges on their nuclei, then the pair of electrons may not be shared equally.

| non-polar covalent bond | polar covalent bond |
| electrons shared equally | unequal sharing of electrons |

Electronegativity is a measure of the attraction an atom has for the electrons in a bond.

The electronegativity values of most elements are given in the Data Booklet (page 11). Note that the values have no units.

Electronegativity increases across a period from left to right. This is because the positive nuclear charge is increasing and therefore the atoms exert a stronger pull on the shared pairs of electrons.

Electronegativity decreases down a group. The reasons for this include:

- as the atomic size increases, the shared pairs of electrons are further away from the nucleus and are attracted less strongly

- there is a greater screening effect because there are more shells of inner electrons between the shared pair of electrons and the nucleus.

ONLINE

Watch the slideshow 'Periodic Trends in Electronegativity' at www.brightredbooks.net

DON'T FORGET

The elements with the highest electronegativities are at the top right corner of the Periodic Table. The elements with the lowest electronegativities are at the bottom left corner of the Periodic Table.

THINGS TO DO AND THINK ABOUT

Explain (i) why there is no fourth ionisation energy value given for lithium and (ii) the large 'jump' between the third and fourth ionisation energy values for aluminium.

ONLINE TEST

Test yourself on periodicity online at www.brightredbooks.net

STRUCTURE AND BONDING 1

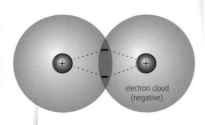

COVALENT AND IONIC BONDS

Small differences in electronegativity values

When two non-metal atoms with identical electronegativity values come together to form a bond, a pure or non-polar covalent bond is formed in which the shared pair of electrons is shared equally between the atoms.

Non-polar
covalent bonds

$$H—H \qquad F—F \qquad Cl—N$$

$$2.2 \quad 2.2 \qquad 4.0 \quad 4.0 \qquad 3.0 \quad 3.0$$

If two non-metal atoms with slightly different electronegativity values come together, a polar covalent bond can be formed.

Polar
covalent bonds

$$\overset{\delta^+}{H}\to\overset{\delta^-}{F} \qquad \overset{\delta^-}{Cl}\leftarrow\overset{\delta^+}{H} \qquad \overset{\delta^-}{F}\leftarrow\overset{\delta^+}{N}$$

$$2.2 \quad 4.0 \qquad 3.0 \quad 2.2 \qquad 4.0 \quad 3.0$$

An arrowhead shown halfway along a bond indicates that the bond is polar. The arrow points towards the more electronegative of the two bonded atoms. The more electronegative atom will carry a slightly negative charge ($\delta-$) and the less electronegative atom will carry a slightly positive charge ($\delta+$).

Large differences in electronegativity values

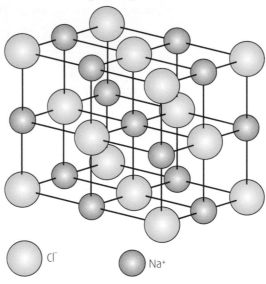

Cl^- Na^+

Ionic compounds form lattice structures. This is Na^+Cl^-.

When the difference in electronegativity values between two atoms is very large, rather than the bonding electrons being shared, the more electronegative atom takes an electron away from the other atom, forming ions.

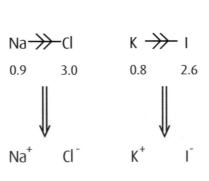

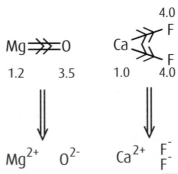

$$Na\ggg Cl \qquad K\ggg I \qquad Mg\ggg O \qquad Ca\overset{4.0}{\underset{4.0}{\overset{F}{\underset{F}{\ggg}}}}$$

$$0.9 \quad 3.0 \qquad 0.8 \quad 2.6 \qquad 1.2 \quad 3.5 \qquad 1.0$$

$$\Downarrow \qquad\qquad \Downarrow \qquad\qquad \Downarrow \qquad\qquad \Downarrow$$

$$Na^+ \quad Cl^- \qquad K^+ \quad I^- \qquad Mg^{2+} \quad O^{2-} \qquad Ca^{2+} \quad \begin{matrix}F^-\\F^-\end{matrix}$$

contd

The bonding continuum

Whether atoms bond to form covalent or ionic bonds depends on several factors, one of which is the difference in electronegativity values between the atoms:

- covalent bonding occurs if the difference in electronegativity values between the elements is small or zero
- ionic bonding occurs if the difference in electronegativity values between the elements is large.

Most chemical bonds are somewhere between these two extremes and it is best to think of ionic and covalent bonding as being at opposite ends of a bonding continuum, with varying degrees of polar covalent bonding lying between the extremes.

	H–H	P–H	C–H	Cl–H	N–H	O–H	F–H	Ca–Cl	Li–Cl	K–F
Difference in electronegativity values	0·0	0·0	0·3	0·8	0·8	1·3	1·8	2·0	2·0	3·2
Compound containing bond	H_2	PH_3	CH_4	HCl	NH_3	H_2O	HF	$CaCl_2$	LiCl	KF
Type of bond	Pure or non-polar covalent		Polar covalent					Ionic		
Can the compound conduct electricity when molten/liquid?	✗	✗	✗	✗	✗	✗	✗	✔	✔	✔

In most cases, the bigger the difference in electronegativity values between the atoms, the more polar the bond and the greater the ionic character.

Compounds formed between metals and non-metals are often ionic **but not always**. Tin tetraiodide is a compound containing a metal and a non-metal which is polar covalent rather than ionic.

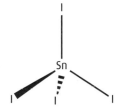

Tin tetraiodide is a compound formed when the metal tin reacts with the non-metal iodine. The electronegativity of tin is 1·8 and the electronegativity of iodine is 2·6, so there is only a difference in electronegativity of 0·8 between the two elements. The bonds formed are polar covalent bonds.

Differences in electronegativity are useful predictors of the type of bonding, but other factors make a contribution and care needs to be taken before jumping to conclusions. **Only by examining the properties of the compound can the type of bonding be identified conclusively.**

The following table shows typical properties of ionic and covalent compounds.

	Ionic	Covalent molecular
State(s) at room temperature	Always solids	Can be solids, liquids or gases
Melting points and boiling points	Usually above 400°C	Usually below 400°C
Electrical conduction when solid	✗	✗
Electrical conduction when liquid	✔	✗
Electrical conduction in solution	✔	✗

ONLINE

Watch the slideshow 'Bonding Continuum' at www.brightredbooks.net

 THINGS TO DO AND THINK ABOUT

1   Using electronegativity values, predict the type of bonding in NaI, CaF_2, SbI_3 and $AuBr_3$.

2   Antimony bromide is a white powder which melts at 97°C forming a colourless liquid that goes on to boil at 280°C. From the properties given, which type of bonding and structure is antimony bromide most likely to have?

 ONLINE TEST

How well have you learned about structure and bonding? Take the test at www.brightredbooks.net

STRUCTURE AND BONDING 2

VAN DER WAALS FORCES

The atoms that make up a molecule are held together by very strong covalent bonds. Between one molecule and its neighbours a far weaker type of 'intermolecular' attraction can exist. At temperatures below the boiling point, these forces can be strong enough to draw molecules towards their neighbours, restricting their movement, and the substance becomes a liquid. At even lower temperatures, below the melting point, these forces will be strong enough to prevent the molecules from moving around and a solid forms.

DON'T FORGET

When covalent molecular substances melt or boil, it is only the weak van der Waals forces **between** molecules that are being broken, not the very strong covalent bonds **inside molecules**.

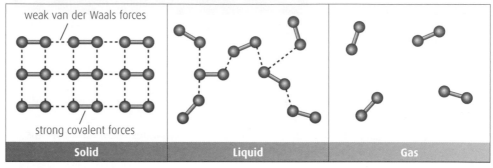

weak van der Waals forces

strong covalent forces

Solid	Liquid	Gas

The name given to any type of force acting between a molecule and neighbouring molecules is a **van der Waals force**. There are three different types of van der Waals forces.

LONDON DISPERSION FORCES

The weakest of all the van der Waals forces are London dispersion forces. London dispersion forces are found in both the monatomic elements (page 12) and the molecular elements (page 14). The strength of London dispersion forces depends on the number of electrons present in each molecule.

Below are the first four members of the family of straight-chain alkanes together with their boiling points:

CH_4	C_2H_6	C_3H_8	C_4H_{10}
$10\,e^-$	$18\,e^-$	$26\,e^-$	$34\,e^-$
$-164°C$	$-89°C$	$-42°C$	$-1°C$

The number of electrons present in a molecule can be worked out using the atomic number of each type of atom present.

DON'T FORGET

The strength of London dispersion forces depends on the total number of electrons present in each molecule.

Example: C_4H_{10}

Number of electrons in $4 \times C$ atoms $= 4 \times 6\,e^- = 24\,e^-$
Number of electrons in $10 \times H$ atoms $= 10 \times 1\,e^- = 10\,e^-$
Total number of electrons in $C_4H_{10} = 34\,e^-$
The greater the number of electrons present in a molecule, the higher the melting and boiling points of the compound will be.

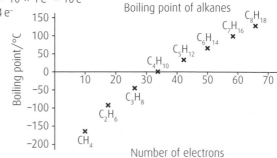

PERMANENT DIPOLE-PERMANENT DIPOLE INTERACTIONS

A **dipole** is the term used to describe molecules in which an unequal distribution of charge results in one side of a molecule becoming slightly positively charged ($\delta+$) while the other side is slightly negatively charged ($\delta-$).

A **polar molecule** is a molecule with a permanent dipole.

For a molecule to be a polar molecule, it must have polar covalent bonds and you must be able to draw a line through the molecule so that on one side of the line all the atoms are slightly positively charged, while on the other side of the line all the atoms are slightly negatively charged.

DON'T FORGET

You must look at both the electronegativity values and the shape of a molecule to decide if it is polar.

Example: 1. Hydrogen sulfide

A line can be drawn through this molecule: the atoms on one side of the line are slightly negatively charged and the atoms on the other side of the line are slightly positively charged. Hydrogen sulfide is a polar molecule.

Example: 2. Carbon dioxide

You cannot draw a line through this molecule in such a way that the atoms on one side are all slightly negatively charged and the atoms on the other side of the line are all slightly positively charged. Carbon dioxide is not a polar molecule.

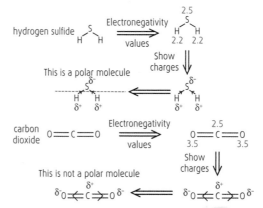

The intermolecular forces between non-polar molecules are the weak London dispersion forces. However, because polar molecules have a permanent dipole, they have stronger intermolecular forces called **permanent dipole–permanent dipole interactions**. These are electrostatic forces of attraction between the permanent dipole of a polar molecule and the permanent dipole of its neighbour. Permanent dipole–permanent dipole interactions are stronger than London dispersion forces for molecules of equivalent size.

VIDEO LINK

Find out more about polar and non-polar molecules by watching the video at www.brightredbooks.net

HYDROGEN BONDING

Hydrogen bonding is the name given to unusually strong permanent dipole–permanent dipole interactions that arise between highly polar molecules.

Nitrogen, oxygen and fluorine atoms are relatively small atoms with electronegativity values greater than hydrogen. The covalent bonds formed by atoms of these elements with hydrogen share the bonding pair of electrons very unequally and so they form very polar covalent bonds. The shape of molecules containing O−H or N−H covalent bonds is also such that they always form polar molecules. The unusually strong permanent dipoles of molecules containing F−H, O−H or N−H bonds gives rise to permanent dipole–permanent dipole interactions that are stronger than other forms of permanent dipole–permanent dipole interactions, but much weaker than covalent bonds. These interactions are known as hydrogen bonds.

DON'T FORGET

You can predict if the permanent dipole–permanent dipole interactions between molecules will be strong enough to be called 'hydrogen bonds' by looking to see if the molecule contains any O−H, N−H or F−H bonds in its structure.

THINGS TO DO AND THINK ABOUT

1   Which of the following molecules is polar?

ammonia   carbon disulfide   hydrogen cyanide   trichloromethane   ethane

2   Which of the following will have hydrogen bonds between their molecules:

ONLINE TEST

How well have you learned about structure and bonding? Take the test at www.brightredbooks.net

STRUCTURE AND BONDING 3

The **physical properties** of covalent molecular substances are largely determined by the type and strength of the van der Waals (intermolecular) forces acting between their molecules.

MELTING AND BOILING POINTS

The melting point and boiling point of a substance depends on the strength of the intermolecular forces between its molecules.

To predict the relative melting point or boiling point of a substance, the two factors to consider are:
- the type of van der Waals bonding present
- the total number of electrons in the molecule.

London dispersion forces	Permanent dipole–permanent dipole	Hydrogen bonding
propane	**methoxymethane**	**ethanol**
total number of electrons = $(3 \times 6) + (8 \times 1) = 26\ e^-$	total number of electrons = $(2 \times 6) + (6 \times 1) + 8 = 26\ e^-$	total number of electrons = $(2 \times 6) + (6 \times 1) + 8 = 26\ e^-$
boiling point –44°C	boiling point –22°C	boiling point +78°C

DON'T FORGET

The strength of London dispersion forces depends on the total number of electrons present in each molecule.

For molecules with similar numbers of electrons, hydrogen bonding is stronger than normal permanent dipole–permanent dipole interactions which, in turn, are stronger than London dispersion forces. In the above examples, each of the compounds has the same number of electrons so would be expected to have London dispersion forces of similar strengths. The differences in their boiling points are due to the other types of van der Waals forces present.

VISCOSITY

Viscosity is a measure of how thick a liquid is or how slow it is to move. Thick, treacle-like liquids have high viscosities, runny liquids have low viscosities. The stronger the intermolecular forces in a substance are, the more slowly the liquid will move when poured and so the greater the viscosity.

Substances with hydrogen bonding will tend to be much more viscous than substances without hydrogen bonding.

DON'T FORGET

Compounds with O–H, N–H or F–H bonds in their structure will have hydrogen bonding between their molecules.

decane   methanol   water   propane-1,2,3-triol

Increasing viscosity

Decane is a non-polar molecule and will only have London dispersion forces between molecules. It has a very low viscosity and is very runny. The other molecules shown all have hydrogen bonding between their molecules, which makes them more viscous. Propane-1,2,3-triol is the most viscous of the molecules because each of its molecules contains three O–H bonds, which make the propane-1,2,3-triol molecules highly polar, resulting in relatively strong intermolecular forces between its molecules.

SOLUBILITY

Substances will tend to be most soluble in solvents with the same type of intermolecular forces as themselves. Compounds with polar molecules will tend to dissolve in polar liquids, whereas compounds with non-polar molecules will tend to dissolve in non-polar liquids.

Highly polar molecules, such as water, can be strongly attracted to positively or negatively charged ions, so many ionic compounds are soluble in water. Ionic compounds are not soluble in non-polar liquids.

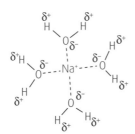

Polar water molecules cluster around the sodium and chloride ions allowing salt to dissolve in water.

DON'T FORGET

Like dissolves like.

UNUSUAL PROPERTIES OF WATER

Water is one of the most common compounds present on our planet and has some remarkable properties. These properties are due to the shape and polarity of the water molecule.

Boiling point

Compared with other molecules containing the same number of electrons, the boiling point of water is unusually high due to the strength of the hydrogen bonding present between its molecules. The hydrogen bonding between water molecules is even stronger than the hydrogen bonding between NH_3 or HF molecules.

Density of ice

For substances other than water, the particles in a solid are packed together far more closely that they would be in a liquid or a gas.

Substances other than water are denser as solids than they are as liquids.

In ice, the intermolecular bonding is hydrogen bonding.

As liquid water cools, the hydrogen bonds begin to lock the molecules together into a type of lattice structure. The hydrogen bonds between the molecules make ice a relatively strong material, but the arrangement of the molecules within the lattice does not allow the molecules to pack closely together. Because of this, ice is less dense than water.

Arrangement of particles in a solid, a liquid and a gas.

Arrangement of water molecules in solid ice.

DON'T FORGET

Water is a good solvent for compounds containing O–H or N–H bonds and will also dissolve many ionic compounds.

10e⁻
b.p. –33°C

10e⁻
b.p. 100°C

10e⁻
b.p. 20°C

DON'T FORGET

Water is unusual because its solid form, ice, is less dense than its liquid form. This is because, in ice, hydrogen bonds lock the water molecules into a lattice of hexagonal rings, leaving lots of empty space between the molecules.

VIDEO LINK

Watch the animation and the video at www.brightredbooks.net to learn more about this topic.

THINGS TO DO AND THINK ABOUT

By thinking about the intermolecular forces present, arrange the following molecules in order from the least viscous to the most viscous:

$CH_3-CH_2-CH_2-CH_3$
butane

$CH_3-CH-CH_2CH_3$ (OH)
butan-2-ol

$CH_3-CH-CH_3$ (O-CH₃)
2-methoxypropane

butane-2,3-diol

ONLINE TEST

How well have you learned about structure and bonding? Take the test at www.brightredbooks.net

NATURE'S CHEMISTRY

ESTERS, FATS AND OILS 1

ESTERS

What are esters?

Esters are a family of organic compounds that are formed when an alcohol reacts with a carboxylic acid. For example, the ester ethyl methanoate is the organic product of the reaction between ethanol (an alcohol) and methanoic acid (a carboxylic acid), as follows:

$$H-\underset{\underset{H}{|}}{\overset{\overset{H}{|}}{C}}-\underset{\underset{H}{|}}{\overset{\overset{H}{|}}{C}}-OH \ + \ HO-\overset{\overset{O}{\|}}{C}-H \ \rightleftharpoons \ H-\underset{\underset{H}{|}}{\overset{\overset{H}{|}}{C}}-\underset{\underset{H}{|}}{\overset{\overset{H}{|}}{C}}-O-\overset{\overset{O}{\|}}{C}-H \ + \ H_2O$$

ethanol   methanoic acid   ethyl methanoate

It is the oxygen and hydrogen atoms coloured red in the above equation that form the water molecule in the reaction. All esters, including ethyl methanoate, contain the following functional group:

$$C-O-\overset{\overset{O}{\|}}{C}-$$

This is known as the **ester group** or **ester linkage** and its presence identifies a compound as an ester. The '**-oate**' ending in the name also identifies a compound as an ester.

DON'T FORGET

A compound can be identified as an ester from its functional group – that is, the ester group – and the '-oate' ending in its name.

Naming esters

Esters are named after the alcohol and carboxylic acid from which they are derived. The first part of the name comes from the name of the alcohol and the second part comes from the name of the carboxylic acid. For example, the ester formed from methanol and ethanoic acid is named methyl ethanoate and the ester formed from ethanol and methanoic acid is called ethyl methanoate.

As well as being able to name an ester if you are given the names of its parent alcohol and carboxylic acid, you must be able to name an ester if you are given its structural formula, such as:

$$H-\underset{\underset{H}{|}}{\overset{\overset{H}{|}}{C}}-\underset{\underset{H}{|}}{\overset{\overset{H}{|}}{C}}-\underset{\underset{H}{|}}{\overset{\overset{H}{|}}{C}}-\overset{\overset{O}{\|}}{C}-O-\underset{\underset{H}{|}}{\overset{\overset{H}{|}}{C}}-\underset{\underset{H}{|}}{\overset{\overset{H}{|}}{C}}-H$$

If we divide the structure into two parts through the C−O bond of the ester linkage, you will notice that the left-hand part contains a C=O group, known as a carbonyl group, and **must** have been derived from the carboxylic acid. It contains four carbon atoms and therefore **butanoic acid** must have been used to make this ester. The other part must have been derived from the alcohol and, as it contains two carbon atoms, the parent alcohol is **ethanol**. So this ester is called **ethyl butanoate**.

Shortened structural formulae can also be written for esters and given these it is possible, although more difficult, to deduce their systematic name, for example **CH₃CH₂CH₂COOCH₃**.

The best way to tackle this is to draw out the full structural formula, making sure the sequence of atoms in the ester group is correct:

$$H-\underset{\underset{H}{|}}{\overset{\overset{H}{|}}{C}}-\underset{\underset{H}{|}}{\overset{\overset{H}{|}}{C}}-\underset{\underset{H}{|}}{\overset{\overset{H}{|}}{C}}-\overset{\overset{O}{\|}}{C}-O-\underset{\underset{H}{|}}{\overset{\overset{H}{|}}{C}}-H$$

We can now see that this ester has been made from methanol and butanoic acid and so it is called **methyl butanoate**.

contd

Structural formulae for esters

Given the name of an ester, you must be able to draw its structural formula. Taking **ethyl propanoate** as an example, one way of doing this is to write the equation for the reaction between its parent alcohol (ethanol) and its parent carboxylic acid (propanoic acid) like that shown on page 24. Another way is to build it up from the ester linkage:

$$C-O-\overset{\displaystyle \overset{O}{\|}}{C}-$$

The part to the right of the dotted line is derived from the carboxylic acid because it contains the carbonyl group (C=O). The parent carboxylic acid is propanoic acid, so we need to attach two more carbon atoms to the carbon atom on the right. One carbon atom also has to be attached to the carbon atom on the left because the parent alcohol is ethanol. We then arrive at the structural formula for ethyl propanoate after adding the required number of hydrogen atoms:

$$H-\overset{\overset{\displaystyle H}{|}}{\underset{\underset{\displaystyle H}{|}}{C}}-\overset{\overset{\displaystyle H}{|}}{\underset{\underset{\displaystyle H}{|}}{C}}-O-\overset{\overset{\displaystyle O}{\|}}{C}-\overset{\overset{\displaystyle H}{|}}{\underset{\underset{\displaystyle H}{|}}{C}}-\overset{\overset{\displaystyle H}{|}}{\underset{\underset{\displaystyle H}{|}}{C}}-H$$

Uses of esters

Many esters have strong, sweet smells and are often floral or fruity in taste. This allows them to be used in the perfume industry as **fragrances** and in the food industry as **flavourings**.

The main use of esters, however, is as industrial **solvents**. Ester molecules are not very polar and, as a result, esters are able to dissolve many compounds which are insoluble in water. Esters also tend to be volatile – they have relatively low boiling points – and this is useful when rapid evaporation of the solvent is required. The ester ethyl ethanoate, for example, is used as the solvent in nail varnish, glues and car body paints, and can also be used to extract caffeine from coffee beans.

DON'T FORGET

Esters are used as fragrances and flavourings, and as industrial solvents.

ONLINE

Revise further how to name esters by following the link at www.brightredbooks.net

ONLINE TEST

Take the 'Esters, fats and oils' test at www.brightredbooks.net

 THINGS TO DO AND THINK ABOUT

1   Name the ester formed between:
   a   butanoic acid and ethanol
   b $CH_3CH_2CH_2CH_2OH$ and CH_3COOH.

2   Name each of the following esters:
   a
$$H-\overset{\overset{\displaystyle H}{|}}{\underset{\underset{\displaystyle H}{|}}{C}}-\overset{\overset{\displaystyle O}{\|}}{C}-O-\overset{\overset{\displaystyle H}{|}}{\underset{\underset{\displaystyle H}{|}}{C}}-\overset{\overset{\displaystyle H}{|}}{\underset{\underset{\displaystyle H}{|}}{C}}-\overset{\overset{\displaystyle H}{|}}{\underset{\underset{\displaystyle H}{|}}{C}}-H$$
   b $CH_3CH_2CH_2CH_2COOCH_2CH_3$
   c $CH_3CH_2CH_2COOCH_2CH_3$

3   Esters and carboxylic acids have the same general formula of $C_nH_{2n}O_2$. This implies that the isomers of a given ester will not only include other esters, but also carboxylic acids. Take ethyl methanoate, for example:

$$H-\overset{\overset{\displaystyle O}{\|}}{C}-O-\overset{\overset{\displaystyle H}{|}}{\underset{\underset{\displaystyle H}{|}}{C}}-\overset{\overset{\displaystyle H}{|}}{\underset{\underset{\displaystyle H}{|}}{C}}-H$$

Draw a structural formula for and write the name of:
   a   an ester that is isomeric with ethyl methanoate
   b   a carboxylic acid that is isomeric with ethyl methanoate.

ESTERS, FATS AND OILS 2

MAKING ESTERS

We learned on page 24 that esters are the organic products of the reaction between carboxylic acids and alcohols. For example, when the carboxylic acid propanoic acid reacts with the alcohol ethanol, the ester ethyl propanoate is formed:

$$H-\underset{\underset{H}{|}}{\overset{\overset{H}{|}}{C}}-\underset{\underset{H}{|}}{\overset{\overset{H}{|}}{C}}-OH \quad + \quad HO-\overset{\overset{O}{\|}}{C}-\underset{\underset{H}{|}}{\overset{\overset{H}{|}}{C}}-\underset{\underset{H}{|}}{\overset{\overset{H}{|}}{C}}-H \quad \rightleftharpoons \quad H-\underset{\underset{H}{|}}{\overset{\overset{H}{|}}{C}}-\underset{\underset{H}{|}}{\overset{\overset{H}{|}}{C}}-O-\overset{\overset{O}{\|}}{C}-\underset{\underset{H}{|}}{\overset{\overset{H}{|}}{C}}-\underset{\underset{H}{|}}{\overset{\overset{H}{|}}{C}}-H \quad + \quad H_2O$$

| ethanol | propanoic acid | ethyl propanoate |

Alternatively, the equation can be written as:

$$H-\underset{\underset{H}{|}}{\overset{\overset{H}{|}}{C}}-\underset{\underset{H}{|}}{\overset{\overset{H}{|}}{C}}-\overset{\overset{O}{\|}}{C}-OH \quad + \quad HO-\underset{\underset{H}{|}}{\overset{\overset{H}{|}}{C}}-\underset{\underset{H}{|}}{\overset{\overset{H}{|}}{C}}-H \quad \rightleftharpoons \quad H-\underset{\underset{H}{|}}{\overset{\overset{H}{|}}{C}}-\underset{\underset{H}{|}}{\overset{\overset{H}{|}}{C}}-\overset{\overset{O}{\|}}{C}-O-\underset{\underset{H}{|}}{\overset{\overset{H}{|}}{C}}-\underset{\underset{H}{|}}{\overset{\overset{H}{|}}{C}}-H \quad + \quad H_2O$$

| propanoic acid | ethanol | ethyl propanoate |

The reactant molecules join through the hydroxyl group of the alcohol and the carboxyl group of the acid with the elimination of a water molecule and the formation of the ester linkage. The hydrogen and oxygen atoms that go to form the water molecule are coloured red in these equations and the groups of atoms which make up the ester linkages are shaded in yellow.

This reaction is described as a **condensation** reaction – that is, a reaction in which two reactant molecules join with the elimination of a small molecule which is usually, but not always, water. The condensation of an alcohol and a carboxylic acid to form an ester can also be referred to as an **esterification** reaction.

Condensation is a **reversible reaction** and, at room temperature, it proceeds at a very slow rate. It can be speeded up by heating the reaction mixture or by adding a catalyst of **concentrated sulfuric acid**. Not only does the concentrated sulfuric acid provide the hydrogen ions needed to catalyse the reaction, it also has a great affinity for water and absorbs the water that is formed in the reaction. This encourages more of the alcohol and carboxylic acid to react, thus increasing the yield of the ester formed.

DON'T FORGET

The ester linkage is formed by the reaction of a hydroxyl group with a carboxyl group.

HYDROLYSIS OF ESTERS

Just as an ester can be made by the condensation reaction between an alcohol and a carboxylic acid, it can also be broken back down into its parent alcohol and carboxylic acid.

Given the name of an ester, you must be able to name the products of its breakdown. We know that the first part of the name of an ester is derived from its parent alcohol and the second part is derived from its parent carboxylic acid. So, for example, the products of the breakdown of the ester **methyl pentanoate** will be **methanol** and **pentanoic acid**.

Given the structural formula for an ester, you must be able to identify the products of the breakdown of that ester. Consider the ester drawn below:

$$H-\underset{\underset{H}{|}}{\overset{\overset{H}{|}}{C}}-\underset{\underset{H}{|}}{\overset{\overset{H}{|}}{C}}-\overset{\overset{O}{\|}}{C}\!+\!O-\underset{\underset{H}{|}}{\overset{\overset{H}{|}}{C}}-\underset{\underset{H}{|}}{\overset{\overset{H}{|}}{C}}-H$$

This ester will break at the point marked by the dotted red line – that is, the C–O bond in the ester linkage. Because the left-hand part of the molecule contains the C=O group, it will form the carboxylic acid, propanoic acid, when the ester breaks down. The right-hand part will form the alcohol, ethanol.

contd

As the condensation reaction that takes place during the formation of an ester and water is reversible, this implies that an ester is broken back down into its parent alcohol and carboxylic acid when it reacts with water. The latter process involves heating the ester with water in the presence of a catalyst such as an acid or alkali. The reaction that takes place is called a **hydrolysis** reaction, for example:

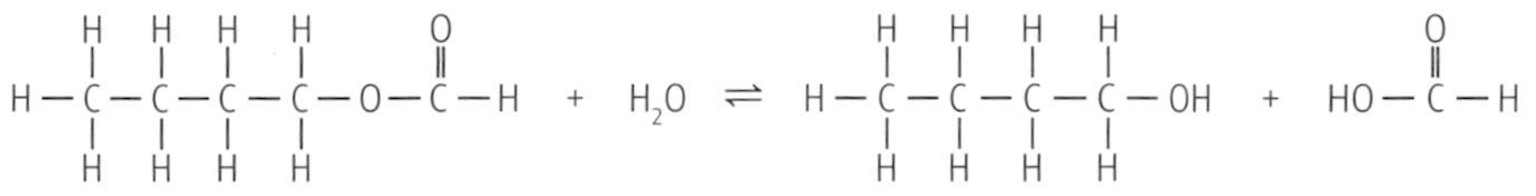

butyl methanoate   butan-1-ol   methanoic acid

In general, a **hydrolysis** reaction is one in which a molecule reacts with water and breaks down into smaller molecules.

Let us now take a closer look at what happens during the hydrolysis of butyl methanoate:

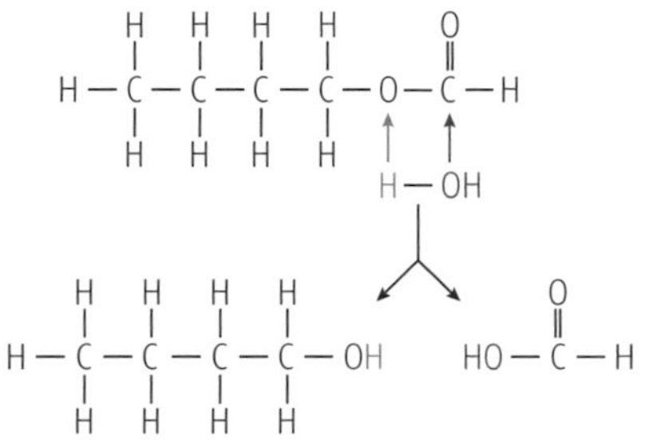

butan-1-ol   methanoic acid

The water molecule attacks the ester linkage in butyl methanoate and breaks the O−C bond (coloured red in the diagram). The −OH group (coloured blue) of the water molecule then joins with the carbon atom of the O−C bond to generate the carboxyl group in the methanoic acid product. The hydrogen atom (coloured green) left over from the water molecule joins with the oxygen atom of the O−C bond to form the hydroxyl group of the alcohol butan-1-ol.

DON'T FORGET

The formation and hydrolysis of an ester are reversible reactions: one is the reverse of the other.

ONLINE

Check out the online tutorial about esters at www.brightredbooks.net

ONLINE TEST

Take the 'Esters, fats and oils' test at www.brightredbooks.net

THINGS TO DO AND THINK ABOUT

Condensation reactions and **dehydration reactions** are often confused because water is a product in both these types of reaction. If we now compare them:

a typical **condensation reaction:**   a typical **dehydration reaction:**

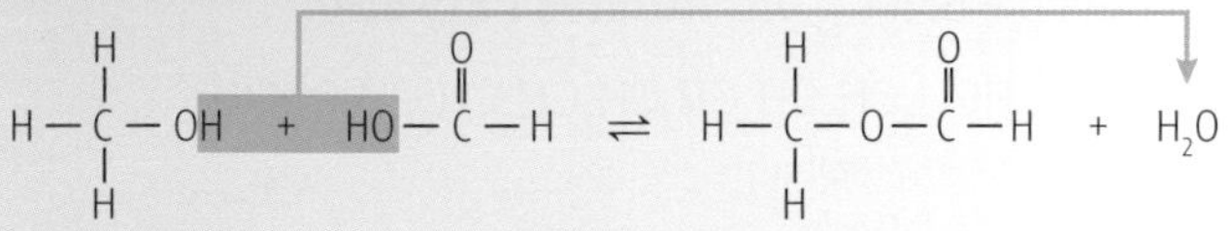

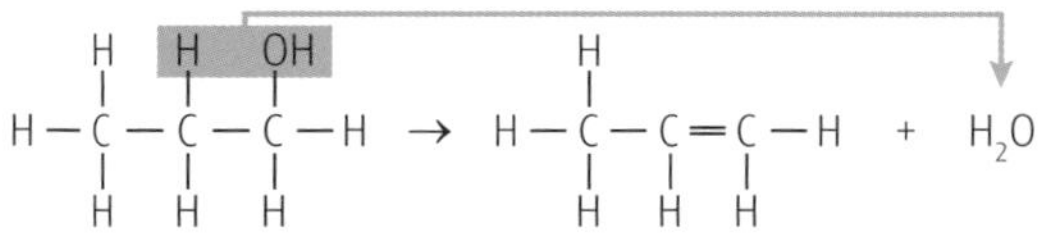

You can see that, in the condensation reaction, the water molecule is removed from **two** reactant molecules, whereas, in the dehydration reaction, the water molecule is taken from just **one** reactant molecule.

Hydrolysis reactions and **hydration reactions** are also often confused as, in both cases, water is a reactant. If we now compare these two types of reaction:

a typical **hydrolysis reaction:**   a typical **hydration reaction:**

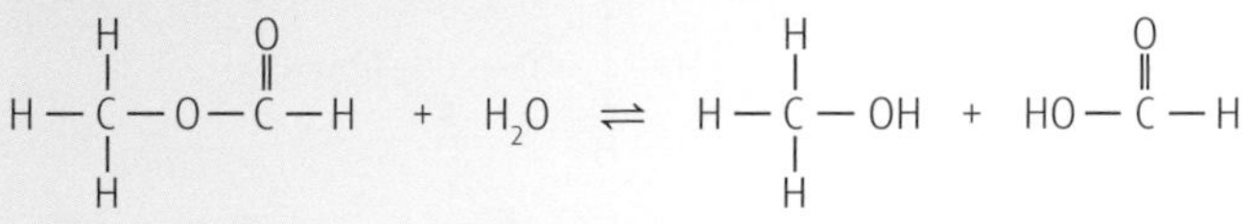

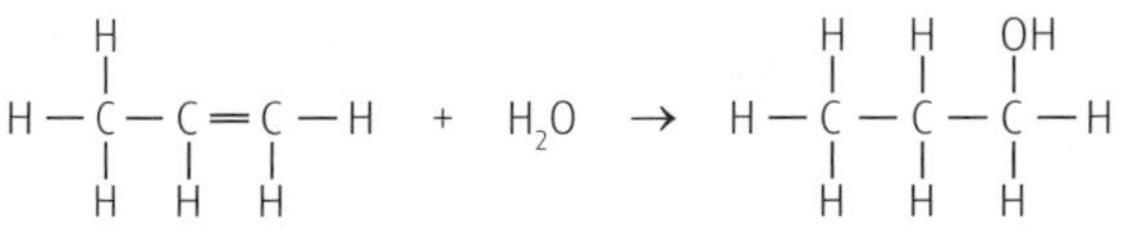

In the hydrolysis reaction, the water breaks down the ester into **two** product molecules, whereas in the hydration reaction, the water adds on to the alkene to form only **one** product molecule.

Condensation and hydrolysis reactions are the reverse of each other and so too are dehydration and hydration reactions.

ESTERS, FATS AND OILS 3

FATS AND OILS

Structure of fats and oils

Edible fats and **edible oils** are naturally occurring compounds and can be obtained from a number of different sources, including:

- **animal sources** – for example, beef fat, butter fat, pork fat (lard)

- **vegetable sources** – for example, olive oil, sunflower oil, linseed oil, rapeseed oil

- **marine sources** – for example, cod liver oil, sardine oil, whale oil.

These are all examples of **esters** and are formed by condensation reactions between the alcohol, **glycerol**, and carboxylic acids known as **fatty acids**.

The structural formula of glycerol is shown below:

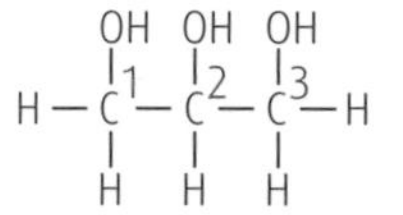

$$
\begin{array}{c}
\text{OH} \quad \text{OH} \quad \text{OH} \\
| \quad 1 \quad | \quad 2 \quad | \quad 3 \\
\text{H} - \text{C} - \text{C} - \text{C} - \text{H} \\
| \qquad | \qquad | \\
\text{H} \quad\;\; \text{H} \quad\;\; \text{H}
\end{array}
$$

glycerol

Glycerol is an alcohol and, as it contains three hydroxyl groups, it is known as a **triol**. It has −OH groups attached at C-1, C-2 and C-3 and its systematic name is **propane-1,2,3-triol**. Note that in naming alcohols with two or more hydroxyl groups the parent alkane name is used in full – that is, we do not drop the 'e' off the end as we do with alcohols containing just one −OH group.

Fatty acids are straight-chain carboxylic acids containing even numbers of carbon atoms ranging from C_4 to C_{24}, although the most common are C_{16} and C_{18}. They can be saturated or unsaturated.

DON'T FORGET

Unsaturated compounds contain at least one C=C bond.

Stearic acid is an example of a C_{18} **saturated fatty acid** and it has the following structural formula:

$$CH_3CH_2CH_2CH_2CH_2CH_2CH_2CH_2CH_2CH_2CH_2CH_2CH_2CH_2CH_2CH_2CH_2COOH$$

stearic acid (saturated)

Oleic acid, on the other hand, is an example of a C_{18} **unsaturated fatty acid**:

$$CH_3CH_2CH_2CH_2CH_2CH_2CH_2CH=CHCH_2CH_2CH_2CH_2CH_2CH_2CH_2CH_2COOH$$

oleic acid (unsaturated)

Because a glycerol molecule contains three hydroxyl groups, it will condense with **three** fatty acid molecules to form the esters present in fats and oils:

glycerol and fatty acids   ester in fat or oil

DON'T FORGET

Fats and oils are esters formed by condensation reactions between glycerol and fatty acids.

$$
\begin{array}{ccc}
 & H & & O & \\
 & | & & \| & \\
 & H-C-OH & HO-C-R & \\
O & | & & H & O \\
\| & & & | & \| \\
R'-C-OH & HO-C-H & \rightleftharpoons & R'-C-O-C-H \\
 & | & & | & \\
 & H-C-OH & HO-C-R'' & + \; 3H_2O \\
 & | & & \| & \\
 & H & & O &
\end{array}
$$

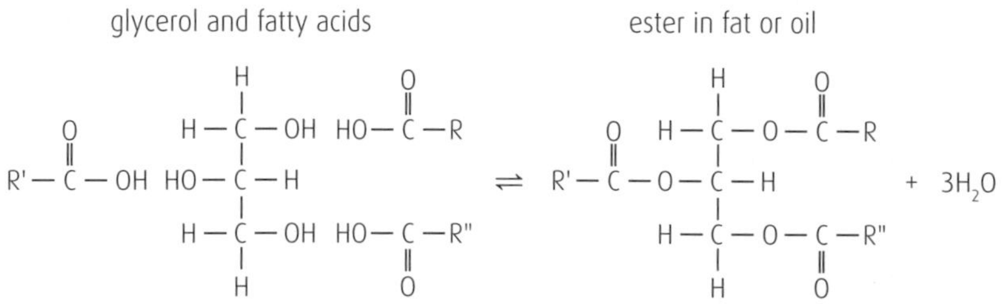

(R, R' and R" represent the hydrocarbon chains of the fatty acids)

An ester formed from glycerol is called a glyceride and so the esters present in fats and oils are referred to as triglycerides. Any fat or oil contains a mixture of triglycerides. In some fats and oils all the fatty acid groups are the same, whereas in others they are different.

contd

Properties of fats and oils

In general, oils decolourise a bromine solution to a much greater extent than fats, indicating that the degree of unsaturation is greater in oils. This difference in the degree of unsaturation accounts for the fact that oils tend to be liquids at room temperature while fats are solids – that is, oils have lower melting points than fats. As oil molecules contain more carbon-to-carbon double bonds than fat molecules, their shapes are considerably different:

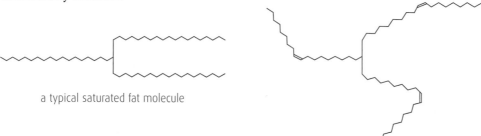

a typical saturated fat molecule

a typical unsaturated oil molecule

Oil molecules are less compact than fat molecules and cannot pack together as closely as fat molecules. This implies that the London dispersion forces between oil molecules are weaker and therefore easier to break than those between fat molecules, thus explaining why oils have lower melting points.

When oils are heated with hydrogen in the presence of a nickel catalyst, the hydrogen molecules add on across some of the carbon-to-carbon double bonds in the oil molecules. The reaction taking place can be described as addition or **hydrogenation**. This reduces the degree of unsaturation in the oil which in turn increases the melting point. In effect, unsaturated liquid oils are converted into saturated solid fats.

Function of fats and oils

Fats and oils are essential components of our diet and play various important roles in our bodies.

Like carbohydrates, fats and oils supply the body with energy, but because the percentage of oxygen in fats and oils is less than that in carbohydrates, they are a much more **concentrated source of energy**. On oxidation, one gram of fat or oil produces about twice as much energy as one gram of carbohydrate. Fats and oils tend to be used as a long-term energy source, whereas carbohydrates are an immediate source of energy.

Fats and oils are needed for the **transport and storage of fat-soluble vitamins** in the body. Vitamins A and D, for example, are fat-soluble vitamins. The reason why these vitamins dissolve in fats and oils is because they are non-polar, just like fats and oils. So a lack of fats and oils in the diet can lead to vitamin deficiencies. A deficiency in vitamin D can result in weakness and deformation of the bones – called rickets – while the consequences of vitamin A deficiency include anaemia, night blindness and growth retardation.

Fats and oils also **supply the body with essential fatty acids**, such as linoleic and linolenic acids, which play an important role in blood coagulation and in brain development.

 THINGS TO DO AND THINK ABOUT

1 Fatty acids are long-chain carboxylic acids and some examples are shown in the following table.

Common name	Systematic name	Structure
stearic acid	octadecanoic acid	$CH_3(CH_2)_{16}COOH$
oleic acid	octadec-9-enoic acid	$CH_3(CH_2)_7CH=CH(CH_2)_7COOH$
linoleic acid	octadec-9,12-dienoic acid	$CH_3(CH_2)_4CH=CHCH_2CH=CH(CH_2)_7COOH$
linolenic acid		$CH_3CH_2CH=CHCH_2CH=CHCH_2CH=CH(CH_2)_7COOH$

What is the systematic name for linolenic acid?

 DON'T FORGET

The lower melting points of oils compared with those of fats are related to the higher degree of unsaturation in oil molecules.

 **VIDEO LINK**

Check out the online tutorial about fats and oils at www.brightredbooks.net

 DON'T FORGET

Fats and oils are a concentrated source of energy and they are essential for the transport and storage of fat-soluble vitamins in the body.

 ONLINE TEST

Take the 'Esters, fats and oils' test at www.brightredbooks.net

 ONLINE

For decades it was believed that there was a link between eating foods containing excessive amounts of saturated fats and heart disease. However, a recent study carried out by a group of researchers at Cambridge University has thrown doubt on such a link. For more information on this, visit the links at www.brightredbooks.net

PROTEINS

FUNCTION OF PROTEINS

Proteins occur in all living cells. In the human body they make up about one-sixth of our body weight. Proteins are the **major structural materials of animal tissue** and give shape to our bodies. They form our hair and nails and, along with water, they are the principal substances of our muscles, organs, blood and skin. Proteins form the collagen of the connective tissue that holds our bones together in a cohesive skeleton and they wrap our bodies in flesh. Proteins are also involved in the **maintenance and regulation of life processes**. Examples of the latter include **enzymes** (the catalysts of biochemical reactions), hormones (such as insulin) and haemoglobin, the oxygen-carrying protein in blood. Whatever their function, all proteins are chemically similar and are composed of the same basic building blocks, called amino acids.

DON'T FORGET

Proteins are the major structural materials of animal tissue and are vital in the maintenance and regulation of life processes.

AMINO ACIDS

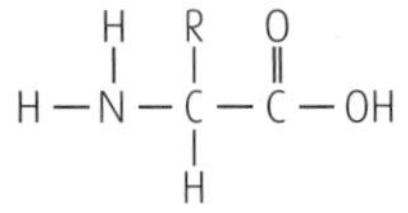

Proteins are large molecules made by linking together lots of **amino acid** molecules. Although about 20 different amino acids are used in synthesising proteins, they can all be represented by the same general structure shown on the left.

They all have in common an **amino group** ($-NH_2$) and a **carboxylic acid group** ($-COOH$) attached to the same carbon atom, but they differ in the R group.

For example, when R is H, the amino acid is called glycine (or aminoethanoic acid) and when R is CH_3, the amino acid is called alanine (or 2-aminopropanoic acid).

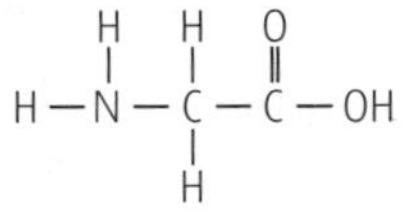

glycine 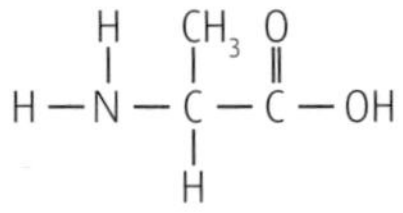   alanine

VIDEO LINK

Learn more about amino acids at www.brightredbooks.net

MAKING PROTEINS

Because amino acids contain both an amino group and a carboxyl group, they are able to condense with each other to form larger molecules. In these **condensation reactions**, the carboxyl group on one amino acid molecule and the amino group on a neighbouring amino acid molecule join together with the elimination of a water molecule. When two amino acids condense together, a dipeptide is formed. For example:

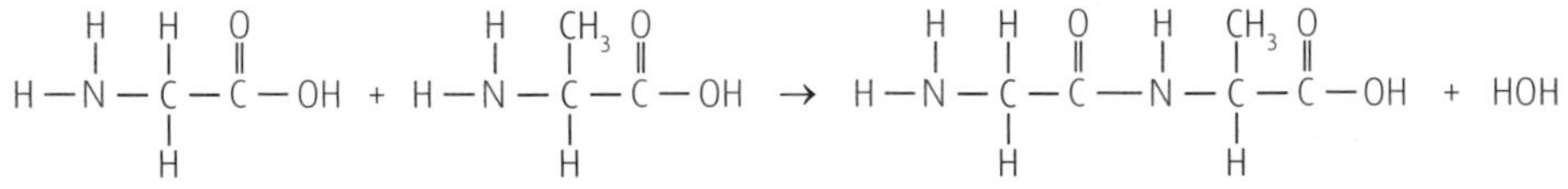

a dipeptide

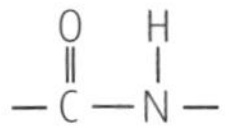

This group of atoms (shown on the left) which links the two amino acids together is called an **amide link**, but when it is present in a peptide it is referred to as a **peptide link**.

When three amino acids condense together, a tripeptide is formed and when a large number link up, a polypeptide is produced:

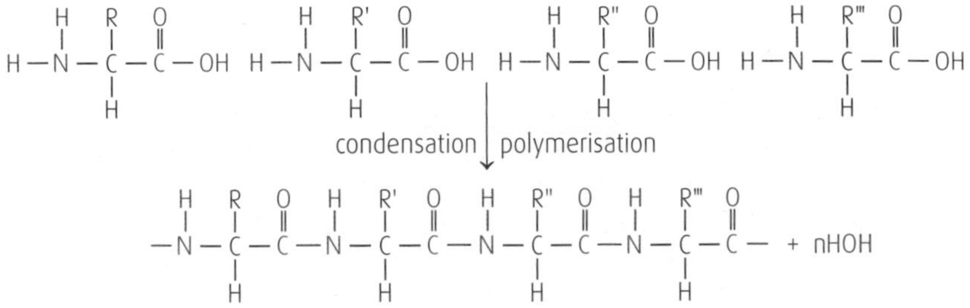

condensation | polymerisation

a polypeptide or protein

contd

Proteins are polypeptides and range in length from about 40 amino acid units to over 4000 amino acid units.

As mentioned earlier, about 20 different amino acids are used to make proteins. So, with 20 different choices available for each amino acid unit in a polypeptide chain, it is not surprising that there are huge numbers of different proteins.

DON'T FORGET

Proteins are large molecules made by linking together individual amino acid units.

HYDROLYSIS OF PROTEINS

During digestion, proteins are broken down into their amino acid units. The proteins react with water and undergo a hydrolysis reaction. This process is catalysed by enzymes. If you are given the structure of a section of a protein, you must be able to draw the structural formulae of the amino acids obtained when it is hydrolysed. For example:

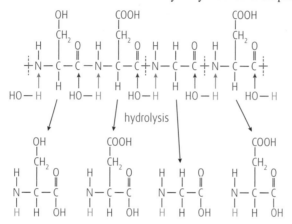

During the hydrolysis reaction, the water molecules attack and break the C−N bonds (coloured red in the diagram) of the peptide links and the individual amino acid molecules are generated as shown above.

ONLINE

Watch an animation showing protein formation at www.brightredbooks.net

DON'T FORGET

Given the structure of a section of a protein, you must be able to draw the structural formulae of the amino acids obtained when it is hydrolysed.

ESSENTIAL AMINO ACIDS

Although plants are able to synthesise all the amino acids they need to make proteins, animals cannot do this. The amino acids that we, as animals, cannot synthesise are known as the **essential amino acids**. We obtain these amino acids by hydrolysing the plant and/or animal proteins in the food we eat. During the digestion of food, these proteins are hydrolysed and the amino acids formed pass into our bloodstream and are carried to various sites in the body where they are reassembled into the specific proteins we need. About half of the amino acids we need are essential amino acids.

DON'T FORGET

Essential amino acids are those amino acids that animals cannot synthesise for themselves.

 THINGS TO DO AND THINK ABOUT

1   The arrangement of amino acids in a peptide is Z-X-W-V-Y, where the letters V, W, X, Y and Z represent amino acids.

On partial hydrolysis of the peptide, which of the following sets of dipeptides is possible?
A   V–Y, Z–X, W–Y, X–W
B   Z–X, V–Y, W–V, X–W
C   Z–X, X–V, W–V, V–Y
D   X–W, X–Z, Z–W, Y–V

2   Each protein has a unique primary structure – that is, the sequence in which the amino acid units are bonded to one another in the protein chain. Insulin was the first protein to be sequenced. This was achieved by Frederick Sanger in 1955. He hydrolysed the insulin and then used chromatography (see page 74) to identify the amino acids present and the order in which they were linked in the chain. This achievement earned him his first Nobel Prize in 1958 and, in 1977, a postage stamp was issued commemorating his work.

ONLINE TEST

Test yourself on proteins online at www.brightredbooks.net

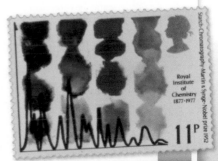

CHEMISTRY OF COOKING 1

FLAVOUR IN FOOD

What is flavour?

The flavours we detect when eating food involve not just our tongues but, more importantly, our noses as well. Our tongues can detect five basic tastes: sweet, sour, bitter, salt and savoury (or umami). The nose detects smells or aromas when the gaseous molecules from volatile compounds trigger receptors in the nose. **Flavour** is therefore a **combination of both taste and smell**. This can be demonstrated by giving someone wearing a blindfold a plain crisp to taste while holding a flavoured crisp under his or her nose. Invariably, the person believes he or she is eating a flavoured rather than a plain crisp.

Properties of flavour compounds

For the molecules of a flavour compound to reach the nose, the compound must be volatile – that is, it must release gaseous molecules into the atmosphere fairly readily. The volatility of a compound will therefore depend on the strength of its intermolecular forces. The weaker the intermolecular forces, the more volatile the compound will be. As intermolecular forces are broken when a compound boils, the boiling point of a compound is a good indicator of its volatility – the lower the boiling point, the more volatile the compound.

Take **limonene**, for example. It is one of the compounds responsible for the flavour in oranges.

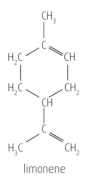
limonene

Limonene molecules contain only hydrogen and carbon and, like other hydrocarbon molecules, they are non-polar. As a result, only London dispersion forces operate between the molecules. As these are the weakest of the van der Waals forces, very little energy is required to break them and this explains why limonene has a relatively low boiling point of 176°C and is volatile.

Let us now consider the flavour compound **vanillin**, found in vanilla pods, and compare it with limonene in an attempt to predict its boiling point and hence its volatility:

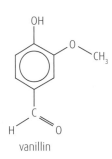
vanillin

Vanillin and limonene, respectively, have 80 and 76 electrons per molecule which are similar enough not to cause any significant difference in their boiling points. However, vanillin contains several functional groups and the most significant of these is the −OH group. The presence of the −OH group implies that hydrogen bonds will exist between neighbouring molecules. The polar nature of the bonds in the other functional groups also suggests the presence of other permanent dipole–permanent dipole attractions. As hydrogen bonds and permanent dipole–permanent dipole interactions are stronger than London dispersion forces, we can predict that the boiling point of vanillin will be greater than that of limonene and vanillin will be less volatile. This is indeed the case – vanillin has a boiling point of 285°C.

Cooking in oil or water

A major issue in cooking is to keep the flavour molecules in the food rather than allowing them to escape into the water or oil in which the food is cooked. Therefore another important property of flavour compounds that is influenced by the presence of functional groups is their **solubility**.

Take asparagusic acid, for example. It is one of the flavour compounds found in asparagus.

Asparagusic acid contains a carboxyl group that can form hydrogen bonds with water molecules, thus making it water-soluble. So, if asparagus is cooked in water, the polar asparagusic acid molecules dissolve in the polar water and are lost when the cooking water is drained away. It therefore makes sense to cook asparagus in oil, as the polar asparagusic acid molecules will not escape into the non-polar oil.

DON'T FORGET

Many of the flavours in foods are due to the presence of volatile compounds. The functional groups present in a flavour molecule influence its volatility.

asparagusic acid

contd

Broccoli, on the other hand, should be cooked in water and the reason for this will become apparent when we consider the structure of allyl isothiocyanate, one of the main flavour compounds in broccoli.

These molecules are not very polar and will be more soluble in non-polar oil than in polar water. So by cooking in water, the flavour compound is retained in the broccoli.

However, one drawback of cooking broccoli in water is the loss of ascorbic acid (vitamin C). The ascorbic acid molecule has lots of sites – the four hydroxyl groups, the oxygen atoms in the carbonyl group and in the ring – which can hydrogen bond with water molecules. It will readily dissolve in water and so be lost from the broccoli.

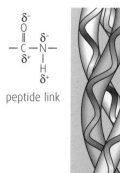

allyl isothiocyanate

ascorbic acid

EFFECT OF HEAT ON PROTEINS

Protein molecules contain polar peptide links along their length and these peptide links can hydrogen bond with others in the same chain to produce a spiral. These spirals can be assembled into more complicated structures which are classified as **fibrous** and **globular**. The **collagen** in meat, for example, is a **fibrous protein**. It consists of three interwoven protein chains forming a rope-like structure (see the diagram on the right). As well as the hydrogen bonds within each chain, hydrogen bonds operate between the chains, making collagen fairly tough. When the meat is cooked, the heat breaks the hydrogen bonds and causes the protein chains to unravel. The collagen is softened and the meat becomes more tender. **Albumin**, the protein in egg white, is an example of a **globular protein**. In globular proteins, the protein chains twist and coil into irregular ball shapes. Heating egg white overcomes the hydrogen bonds that hold the protein in its globular form and allows the chain to uncoil. The uncoiled protein chain then forms hydrogen bonds with other uncoiled molecules and they mesh together to give the familiar white solid of cooked eggs.

In general, when a protein is heated, intermolecular bonds are broken and the shape of the protein is irretrievably changed. When this happens, the protein is said to be **denatured**.

peptide link

 DON'T FORGET

When proteins are heated during cooking, intermolecular bonds are broken and the proteins are denatured.

 VIDEO LINK

Watch the video about the flavour in our food at www.brightredbooks.net

VIDEO LINK

Watch videos showing interesting recipes based on chemical reactions at www.brightredbooks.net

 THINGS TO DO AND THINK ABOUT

1   Vanillin and zingerone are flavour molecules:

vanillin

zingerone

Which line in the table correctly compares the properties of vanillin and zingerone?

	More soluble in water	More volatile
A	vanillin	vanillin
B	vanillin	zingerone
C	zingerone	vanillin
D	zingerone	zingerone

2   When a protein is denatured

   A   its overall shape is distorted

   B   its amide links are hydrolysed

   C   it is broken into separate peptide fragments

   D   it decomposes into amino acids.

 ONLINE TEST

Head to www.brightredbooks.net and take the test on the chemistry of cooking.

CHEMISTRY OF COOKING 2

ALDEHYDES AND KETONES

Structures of aldehydes and ketones

Many flavour compounds are aldehydes and ketones. For example, vanillin is an aldehyde and carvone, which has a spearmint flavour, is a ketone.

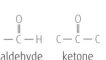

vanillin

carvone

As you can see from these structures, **aldehydes** and **ketones** are closely related families of organic compounds and both contain the **carbonyl functional group** ($>$C=O). Although they share the same functional group, there is a subtle difference in their structures.

In aldehydes, a hydrogen atom is always bonded to the carbonyl group; however, in ketones, the carbonyl group is flanked by two carbon atoms (see the diagrams on the left).

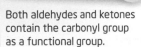

aldehyde   ketone

DON'T FORGET

Both aldehydes and ketones contain the carbonyl group as a functional group.

DON'T FORGET

The abbreviated form of the aldehyde group is –CHO and not –COH.

Naming aldehydes and ketones

The simplest member of the aldehyde series contains only one carbon atom and is known as methanal. The simplest ketone contains three carbon atoms and is known as propanone. Aldehydes and ketones can therefore be identified by their '**–al**' and '**–one**' name endings, respectively.

Let us consider the following branched-chain aldehyde and name it:

$CH_3CH(CH_3)CH_2CHO$

The longest carbon chain containing the carbonyl group has four carbon atoms in it and so the parent name is **butanal**. The carbon atoms are numbered starting at the end nearer the functional group – that is, the right-hand end. A methyl group is attached at C-3 and so the systematic name is **3-methylbutanal**.

You will notice that a number is not used in the name to indicate the position of the functional group. It is not necessary because the carbonyl group in an aldehyde is always at the end of the carbon chain.

Let us now consider the branched-chain ketone shown in the box on the left.

$CH_3COCH(CH_3)CH_2CH_3$

The longest carbon chain containing the carbonyl group contains five carbon atoms and so the parent name is **pentanone**. Numbering starts from the left-hand end as the carbonyl functional group is nearer that end. The carbonyl group is at C-2 and so the parent name becomes **pentan-2-one**. There is a methyl group attached at C-3 and so the systematic name is **3-methylpentan-2-one**.

Drawing structural formulae

As well as being able to name straight-chain and branched-chain aldehydes and ketones if you are given their structural formulae, you need to be able to draw a structural formula for an aldehyde or ketone if you are given its systematic name.

Take **4,4-dimethylhexan-2-one**, for example. We recognise this as a ketone from the **–one** name ending. The parent name is **hexan-2-one**, so we can draw a chain of six carbon atoms with the carbonyl group at C-2. We then attach two methyl groups at C-4 and finally add hydrogen atoms to obtain the structure shown in the second box on the left, with the corresponding shortened structural formulae.

$CH_3COCH_2C(CH_3)_2CH_2CH_3$

contd

Oxidising carbonyl compounds

When aldehydes and ketones burn in a plentiful supply of oxygen, they undergo complete combustion and are fully oxidised to carbon dioxide and water. Under less severe conditions, some carbonyl compounds can be partially oxidised to new organic compounds in which the carbon skeleton remains intact.

As a result of their structural differences, aldehydes can be oxidised to **carboxylic acids**, but ketones resist mild oxidation. For example:

You will notice that, when an aldehyde is oxidised, an oxygen atom is inserted into the C–H bond attached to the carbonyl group. A ketone does not have a hydrogen atom attached to the carbonyl group and this accounts for the fact that it resists mild oxidation.

The following table shows the oxidising agents that can be used to oxidise aldehydes to carboxylic acids, together with the results observed.

DON'T FORGET

Aldehydes can be oxidised to carboxylic acids, but ketones resist mild oxidation.

Oxidising agents and conditions	Observations
hot copper(II) oxide	**black** copper(II) oxide reduced to **brown** copper
warm acidified potassium dichromate solution	**orange** dichromate ions reduced to **blue–green** chromium(III) ions
warm Fehling's solution (alkaline solution containing $Cu^{2+}(aq)$ ions)	**blue** copper(II) ions reduced to a **brick red** precipitate of copper(I) oxide
warm Tollens' reagent (alkaline solution containing $Ag^+(aq)$ ions)	**colourless** silver(I) ions are reduced to a **grey** solid (a silver mirror)

As ketones cannot be oxidised, these reagents can be used to distinguish aldehydes from ketones.

We usually describe oxidation in terms of a loss of electrons, but, when applied to organic compounds, **oxidation** results in **an increase in the oxygen to hydrogen (O:H) ratio**. Consider, for example, the conversion of ethanal to ethanoic acid:

$$CH_3CHO \rightarrow CH_3COOH$$

In ethanal, the O:H ratio is 1:4, whereas in ethanoic acid it is 2:4. The O:H ratio has increased and so this reaction is an oxidation reaction.

DON'T FORGET

When an organic compound is oxidised, its O:H ratio increases.

VIDEO LINK

To see a spectacular demonstration of the Tollens' reagent test, visit www.brightredbooks.net

THINGS TO DO AND THINK ABOUT

Not only can **Fehling's solution** and **Tollens' reagent** be used to distinguish aldehydes from ketones, they can also be used to **identify** an organic compound as an **aldehyde**. In other words, if an organic compound gives a positive test with Fehling's solution or Tollens' reagent, then that organic compound **is** an aldehyde.

Another interesting fact about Tollens' reagent was its earlier use in the manufacture of **mirrors**. A plate of glass was exposed to a mixture of Tollens' reagent and glucose.

ONLINE

Read more about naming organic compounds by following the link at www.brightredbooks.net

Glucose has a ring structure and, in aqueous solution, the ring opens and an equilibrium is established between the ring structure and the open-chain structure. You can see that the latter contains an aldehyde group (shown in green) and this allowed it to reduce the silver(I) ions in the Tollens' reagent to silver metal, which was then deposited as a thin layer on the surface of the glass plate.

glucose (ring structure) ⇌ glucose (open-chain structure)

ONLINE TEST

Head to www.brightredbooks.net and take the test on the chemistry of cooking.

OXIDATION OF FOOD 1

DON'T FORGET

As well as being able to name straight-chain and branched-chain alcohols if you are given their structural formulae, you need to be able to draw a structural formula for an alcohol given its systematic name.

EXPOSING FOOD TO THE AIR

When food is exposed to the air, many of the chemicals in the food react with oxygen and the food is 'spoiled'. For example, when an apple is cut, the exposed surface slowly turns brown as it reacts with oxygen and edible oils turn rancid on long exposure to the air. In the latter case, the oxygen reacts with the unsaturated triglycerides, breaking them down into foul-smelling molecules.

The chemicals in food undergo **oxidation reactions** on exposure to oxygen. There are a number of ways of keeping food fresh. For example, refrigerating food will slow down these oxidation reactions. Packaging food in an atmosphere of nitrogen rather than air will extend its shelf life. Adding antioxidants (see page 39) to food stops it from spoiling because they prevent the food molecules from being oxidised.

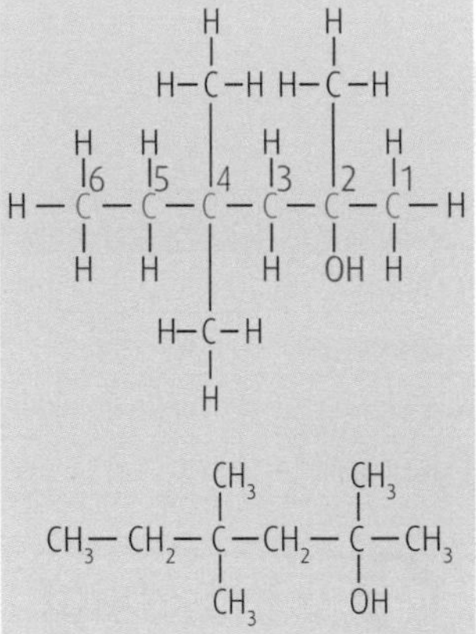

ALCOHOLS

Naming alcohols

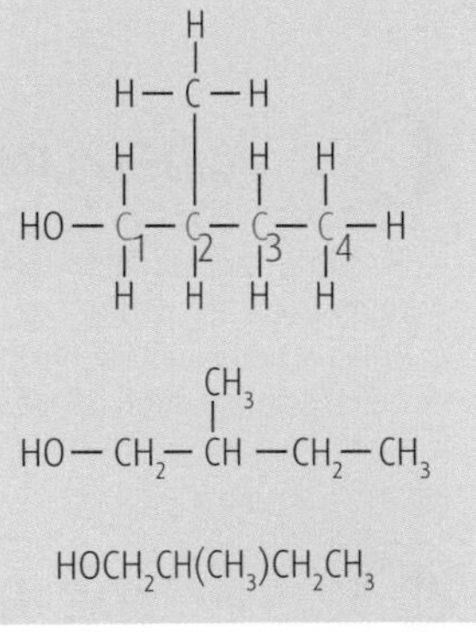

Alcohols are a family of organic compounds in which the **hydroxyl group** or **−OH group** is the functional group and they all have names ending in '**–ol**'.

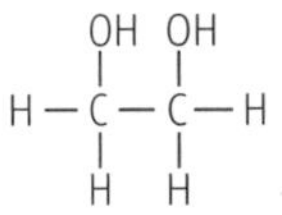

Let us consider the branched-chain alcohol in the box on the left and name it.

The longest carbon chain to which the −OH group is attached contains six carbon atoms, which means the parent name is **hexanol**. Numbering starts from the right-hand end because the −OH group is nearer that end. The −OH group is attached at C-2 and so the parent name becomes **hexan-2-ol**. There are three branches and all are methyl groups – one is attached to C-2 and two are attached to C-4. The systematic name for this alcohol is therefore **2,4,4-trimethylhexan-2-ol**.

If you are given the systematic name of an alcohol, you must be able to draw its structural formula.

Take **2-methylbutan-1-ol**, for example. The parent name is **butan-1-ol**, so we can draw a chain of four carbon atoms with the hydroxyl group on C-1. We then attach a methyl group at C-2 and finally add hydrogen atoms to obtain the structure shown in the box on the left together with the corresponding shortened structural formulae.

Alcohols with more than one −OH group

We have already come across an alcohol with more than one −OH group when we considered fats and oils (see page 28). This was glycerol and it is an example of a **triol** because it contains three hydroxyl groups (see structure on the left).

You will recall that its systematic name is **propane-1,2,3-triol**. Remember when naming triols that the parent alkane name is used in full and the numbers of the carbon atoms to which the hydroxyl groups are attached are also shown in the name.

Consider now the **diol** commonly known as ethylene glycol, which is the main constituent of antifreeze. It contains two carbon atoms and so the parent name is ethane. One −OH group is attached to C-1 and the other to C-2. Its systematic name is therefore **ethane-1,2-diol**.

Physical properties of alcohols

Alcohols have significantly **higher boiling points** than alkanes with similar numbers of electrons per molecule. The reason is the presence of the **polar −OH group**, which allows **hydrogen bonds** to form between the alcohol molecules. This arrangement is illustrated in the diagram on the top right of p37.

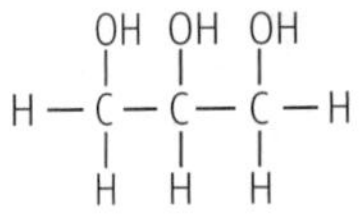

glycerol

ethylene glycol

contd

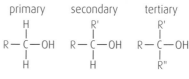

R represents an alkyl group and ▬▬▬ represents hydrogen bonding

Hydrogen bonds are stronger than the London dispersion forces that operate between the alkane molecules. More energy is therefore needed to break the hydrogen bonds, thus accounting for the higher boiling points of the alcohols.

The **polar −OH group** in alcohols also allows them to form hydrogen bonds with **polar water molecules** and this explains their **solubility** in water.

Oxidising alcohols

Like aldehydes, some alcohols can be oxidised, but the structure of the alcohol has an important bearing on the outcome of the oxidation process. There are three structural types of alcohol: **primary**, **secondary** and **tertiary**. The classification depends on the number of alkyl groups that are attached to the hydroxyl-bearing carbon atom.

Primary alcohols have **one** alkyl group attached to the hydroxyl-bearing carbon atom. Secondary alcohols have **two** alkyl groups attached to the hydroxyl-bearing carbon atom and tertiary alcohols have **three**.

Despite the fact that it has no alkyl group attached to the hydroxyl-bearing carbon atom, **methanol** is classified as a primary alcohol because it undergoes similar chemical reactions to the other primary alcohols.

Primary alcohols are oxidised to **aldehydes**, secondary alcohols are oxidised to **ketones** and tertiary alcohols resist mild oxidation. This is summarised below using the three isomeric alcohols butan-1-ol (primary), butan-2-ol (secondary) and 2-methylpropan-2-ol (tertiary):

(R, R' and R" represent alkyl groups)

methanol

You will notice that when oxidation does take place, two hydrogen atoms (shaded blue) are removed from the alcohol, one from the −OH group and one from the hydroxyl-bearing carbon atom. The tertiary alcohol has no hydrogen atom directly attached to the hydroxyl-bearing carbon atom and this is why it resists mild oxidation.

 DON'T FORGET

You must be able to identify an alcohol as primary, secondary or tertiary given its structural formula and remember that methanol is an example of a primary alcohol.

 DON'T FORGET

Primary and secondary alcohols are oxidised to aldehydes and ketones, respectively, whereas tertiary alcohols resist mild oxidation.

The following table shows the oxidising agents that can be used to oxidise alcohols to aldehydes and ketones, together with the results observed.

Oxidising agents and conditions	Observations
hot copper(II) oxide	**black** copper(II) oxide reduced to **brown** copper
warm acidified potassium dichromate solution	**orange** dichromate ions reduced to **blue-green** chromium(III) ions

You will notice that the oxidising agents Fehling's solution and Tollens' reagent do not feature in this table. They can only oxidise aldehydes – they are not strong enough to oxidise primary and secondary alcohols.

THINGS TO DO AND THINK ABOUT

1 Compound **X** reacts with hot copper(II) oxide and the organic product did **not** produce a colour change with Fehling's solution. Compound **X** could be:
   A butan-1-ol   B butan-2-ol   C butanone   D butanoic acid.

2 For centuries, alcoholic beverages have been consumed by humans mainly for recreational reasons. The alcohol in these drinks is ethanol and it is broken down in the body by a series of **oxidation** reactions catalysed by enzymes. In the liver, ethanol is oxidised to ethanal, which, in turn, is oxidised to ethanoic acid. The latter is finally oxidised to carbon dioxide and water at various sites in the body. Ethanal is the most toxic of these chemicals and it is believed to be the main cause of liver damage (cirrhosis). Recent studies have suggested that ethanal may also play a role in the development of alcohol addiction.

 ONLINE

Learn more about alcohols by following the link at www.brightredbooks.net

ONLINE TEST

How well have you learned about the oxidation of food? Take the test at www.brightredbooks.net

OXIDATION OF FOOD 2

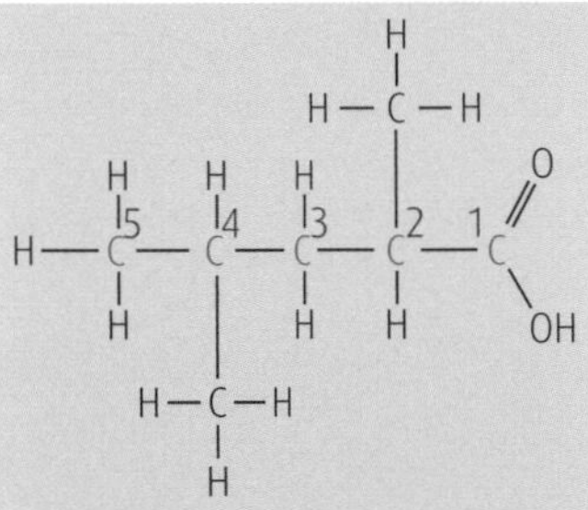

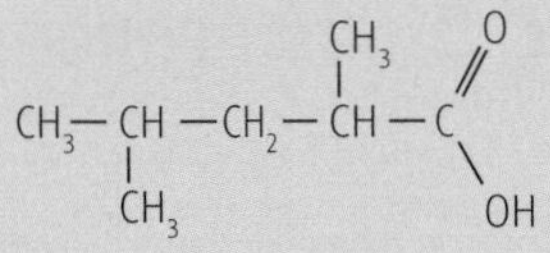

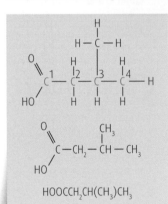

CH₃CH(CH₃)CH₂CH(CH₃)COOH

HOOCCH₂CH(CH₃)CH₃

CARBOXYLIC ACIDS

Naming carboxylic acids

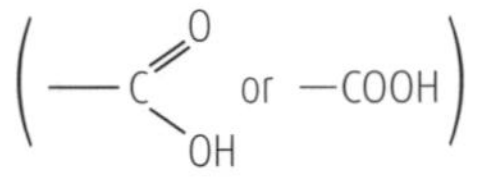

Carboxylic acids are a family of organic compounds containing the **carboxyl group** as the functional group (see diagram on the left).

We can see that the carboxyl group is made up of a carbonyl group and a hydroxyl group, but both are attached to each other. This arrangement modifies the characteristic behaviours of the carbonyl and hydroxyl groups and this is why the carboxyl group is considered a functional group in its own right.

Carboxylic acids are named from the parent alkane by omitting the final '–e' and adding the ending '–oic acid'.

Let us consider the branched-chain carboxylic acid shown in the box on the left and name it.

The longest carbon chain containing the functional group has five carbon atoms in it, so the parent name is **pentanoic acid**. Numbering starts from the right-hand end as the functional group is at that end. There are two branches and both are methyl groups – one is attached to C-2 and the other to C-4. So this carboxylic acid is called **2,4-dimethylpentanoic acid**.

As well as being able to name straight-chain and branched-chain carboxylic acids if you are given their structural formulae, you need to be able draw a structural formula for a carboxylic acid if you are given its systematic name.

Take **3-methylbutanoic acid**, for example. The parent name is **butanoic acid**, so we can draw a chain of four carbon atoms with the carboxyl group at C-1. We then attach a methyl group at C-3 and finally add hydrogen atoms to obtain the structure shown in the box on the left with the corresponding shortened structural formulae.

Isomers of carboxylic acids

You will recall that **isomers** are compounds with the **same molecular formula**, but **different structural formulae**.

Let us consider the carboxylic acids with the molecular formula $C_4H_8O_2$. There are two such carboxylic acids and their structures and names are shown below:

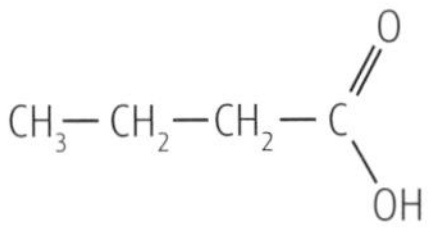

$$CH_3-CH_2-CH_2-C{\overset{O}{\underset{OH}{}}}$$

butanoic acid

$$CH_3-CH-C{\overset{O}{\underset{OH}{}}}$$ with CH₃ branch

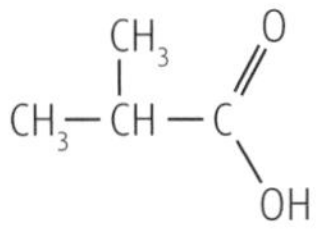

2-methylpropanoic acid

Esters, however, share the same general formula as carboxylic acids and there are four esters with the molecular formula $C_4H_8O_2$. Their structures and the names of three of them are shown below:

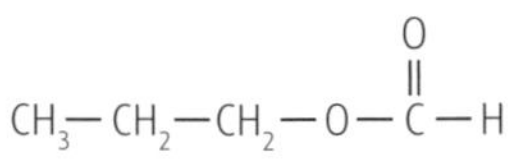

$$CH_3-CH_2-CH_2-O-\overset{O}{\overset{\|}{C}}-H$$

propyl methanoate

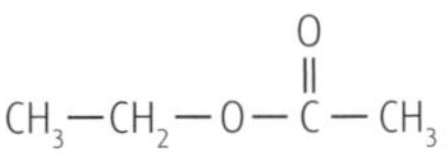

$$CH_3-CH_2-O-\overset{O}{\overset{\|}{C}}-CH_3$$

ethyl ethanoate

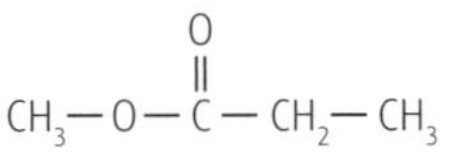

$$CH_3-O-\overset{O}{\overset{\|}{C}}-CH_2-CH_3$$

methyl propanoate

$$CH_3-CH-O-\overset{O}{\overset{\|}{C}}-H$$ with CH₃ branch

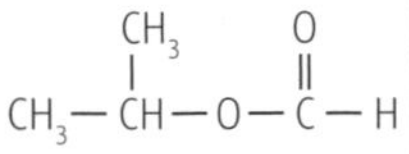

So there are, in total, six isomers with the molecular formula $C_4H_8O_2$.

Reduction of carboxylic acids

We learned on pages 37 and 35 that primary alcohols can be oxidised to aldehydes which can, in turn, be oxidised to carboxylic acids, and that secondary alcohols can be oxidised to ketones. The reverse processes (**reductions**) can also be brought about – that is,

contd

carboxylic acids can be reduced to aldehydes which can, in turn, be reduced to primary alcohols, and ketones can be reduced to secondary alcohols. These reduction reactions are summarised on the right.

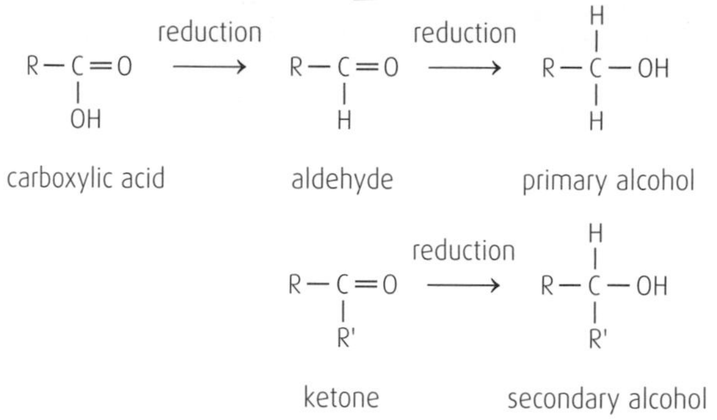

We also learned on page 35 that, when applied to organic compounds, oxidation results in an increase in the O:H ratio. Similarly, **reduction** will result in a **decrease in the O:H ratio**.

Consider the following reaction, for example:

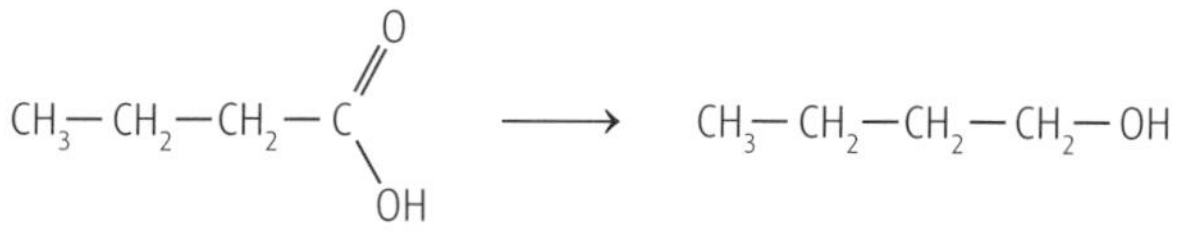

$$CH_3-CH_2-CH_2-C\overset{O}{\underset{OH}{}} \longrightarrow CH_3-CH_2-CH_2-CH_2-OH$$

butanoic acid   butan-1-ol

In butanoic acid, the O:H ratio is 2:8 (0·25), whereas in butan-1-ol it is 1:10 (0·1). The O:H ratio has decreased and so this reaction is a reduction.

Other reactions of carboxylic acids

In aqueous solution, **carboxylic acids** behave as typical acids and form salts on reaction with:
- some **metals** – for example, $Mg + 2CH_3COOH \rightarrow H_2 + Mg^{2+}(CH_3COO^-)_2$
- **alkalis** – for example, $KOH + CH_3CH_2COOH \rightarrow H_2O + K^+CH_3CH_2COO^-$
- **carbonates** – for example, $Na_2CO_3 + 2HCOOH \rightarrow H_2O + CO_2 + 2Na^+HCOO^-$
- **metal oxides** – for example, $CuO + 2CH_3COOH \rightarrow H_2O + Cu^{2+}(CH_3COO^-)_2$.

The first of these is a redox reaction and the other three are neutralisation reactions.

DON'T FORGET

When an organic compound is reduced, its O:H ratio decreases.

ANTIOXIDANTS

Food 'spoils' on exposure to air because compounds in the food react with oxygen – that is, they undergo oxidation reactions. To prevent this happening, **antioxidants** can be added to food. An antioxidant is a compound that it is so easily oxidised itself that it protects other compounds from oxidation. In other words, the antioxidant is preferentially oxidised and therefore sacrificed to protect the food compounds.

Vitamin C (ascorbic acid), a water-soluble compound, is one of the most common antioxidants added to foods. It readily undergoes oxidation and the process is summarised in the following ion–electron equation:

$$C_6H_8O_6(aq) \rightarrow C_6H_6O_6(aq) + 2H^+(aq) + 2e^-$$

You will learn more about writing ion–electron equations like this on page 72.

To prevent the oxidation of edible fats and oils, the synthetic antioxidants butylated hydroxyanisole (BHA) and butylated hydroxytoluene (BHT) and the natural antioxidant vitamin E are used in foodstuffs. These antioxidants are appropriate for this use as they are soluble in fats and oils. Their structures are drawn below:

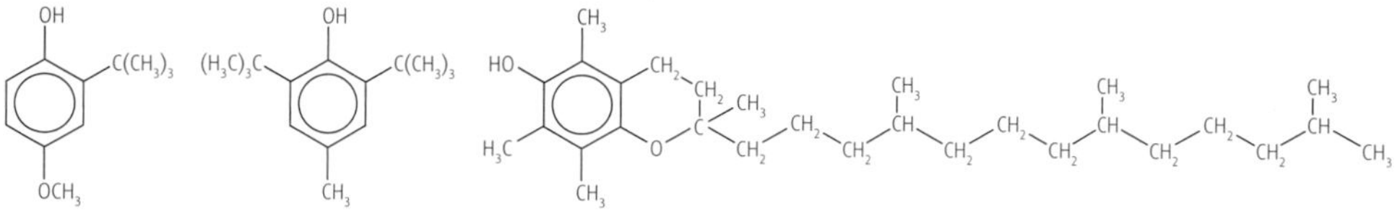

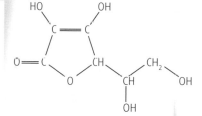
vitamin C
(ascorbic acid)

BHA   BHT   vitamin E

DON'T FORGET

Antioxidants are used to prevent food compounds from being oxidised on exposure to air. The antioxidants are oxidised instead.

 VIDEO LINK

Watch the clip introducing carboxylic acids at www.brightredbooks.net

 ## THINGS TO DO AND THINK ABOUT

Malonic acid has many uses in the pharmaceutical, cosmetic and food-processing industries. It is a dicarboxylic acid with the molecular formula $C_3H_4O_4$. Draw a structural formula for malonic acid.

 ONLINE TEST

How well have you learned about the oxidation of food? Take the test at www.brightredbooks.net

SOAPS, DETERGENTS AND EMULSIONS

SOAPS

Making soaps

Just as a simple ester can undergo **hydrolysis** (reaction with water) to form an alcohol and carboxylic acid, fats and oils can also be hydrolysed to produce glycerol and fatty acids:

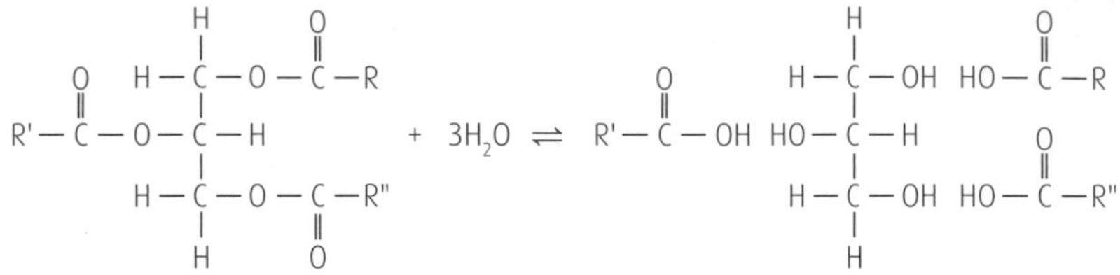

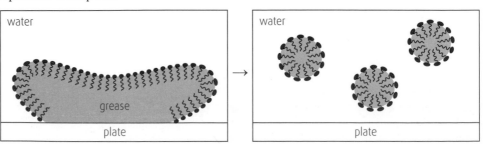

Notice that for every one mole of fat or oil that is hydrolysed, one mole of glycerol and three moles of fatty acids are formed.

It is this hydrolysis reaction that forms the basis of **soap-making**. In the manufacture of soaps, fats and oils are boiled with sodium (or potassium) hydroxide solution. The alkali first catalyses the hydrolysis reaction and then neutralises the fatty acids to form their sodium (or potassium) salts:

$$R'-\overset{\overset{O}{\|}}{C}-OH \ + \ NaOH \longrightarrow R'-\overset{\overset{O}{\|}}{C}-O^-Na^+ \ + \ H_2O$$

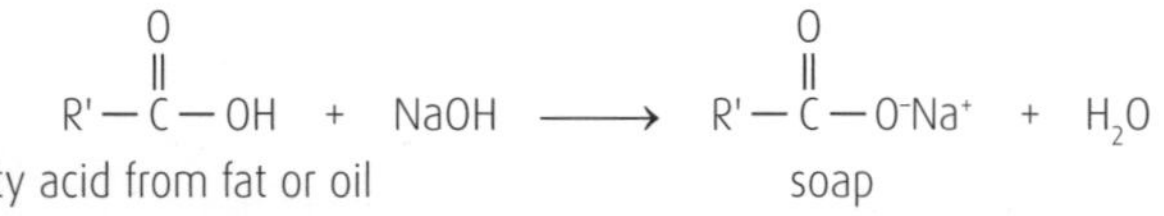

fatty acid from fat or oil   soap

It is these **sodium (or potassium) salts of fatty acids** that are **soaps**.

Cleansing action of soaps

Examples of soaps include sodium stearate, $CH_3(CH_2)_{16}COO^-Na^+$, and potassium palmitate, $CH_3(CH_2)_{14}COO^-K^+$. Their cleansing action can be explained in terms of the structure and bonding of their negative ions. Take the stearate ion, for example:

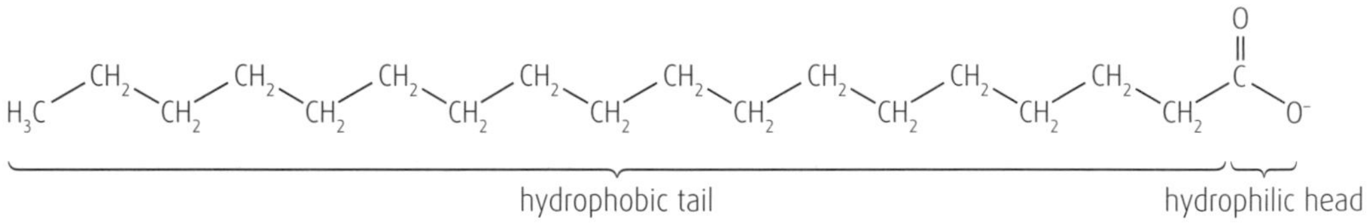

hydrophobic tail   hydrophilic head

The **hydrophilic head** of the soap ion is **soluble in water** because it is **ionic**. The long **hydrocarbon tail** is **non-polar** and is **insoluble in water** (**hydrophobic**), but it is **soluble** in organic materials, such as **oil or grease**. When soap is used to clean greasy plates, for example, its hydrophobic tails bury themselves in the grease while the hydrophilic heads remain in the water. The soap ions dislodge the grease from the surface of the plates and the grease splits up into tiny droplets which are dispersed in the water, leaving the plates clean. This process is illustrated in the following diagram, where 〰● represents a soap ion:

As the head of each soap ion is negatively charged, the grease droplets repel each other, thus preventing them from joining together again. The grease is effectively dispersed through the water. Without the soap, the grease and water would not mix.

DETERGENTS

Detergents can be found in a variety of household cleaning products, including washing powders, washing-up liquids, carpet cleaners and shampoos. They are synthetic and usually derived from crude oil. One of the most common synthetic detergents in use today is sodium dodecylbenzenesulfonate, $CH_3(CH_2)_{11}C_6H_4SO_3^-Na^+$. The negative ion of this salt has the structure shown on the right.

hydrophobic tail   hydrophilic

Like the soap ion, the detergent ion has a long, non-polar hydrocarbon tail (hydrophobic) and an ionic head (hydrophilic). This means that the cleansing action of detergent ions is identical to that of soap ions.

One advantage that detergents have over soaps, however, is their use in hard water areas. Water is described as being hard when it contains $Ca^{2+}(aq)$ and/or $Mg^{2+}(aq)$. These ions react with soap ions to form insoluble salts known as scum. This reduces the efficiency of soaps as cleansing agents. However, when detergents are used, no scum is formed because the calcium and magnesium salts of detergents are soluble in water.

DON'T FORGET

The cleansing action of both soap and detergent ions can be explained in terms of their hydrophilic heads and hydrophobic tails.

EMULSIONS

When two immiscible liquids – that is, liquids which do not dissolve in each other – are added together and vigorously shaken, an emulsion is formed. An **emulsion** is therefore a **mixture** and not a solution and consists of one of the liquids dispersed throughout the other. Vegetable oils and water are often used in emulsions and if such emulsions are left to stand, the oil and water separate into two distinct layers. To prevent this happening, **emulsifiers** are added. They are compounds similar in structure to soap and detergent ions in that they have hydrophobic tails and hydrophilic heads. The structures of two typical emulsifiers are illustrated below:

VIDEO LINK

Watch a short clip introducing emulsifiers at www.brightredbooks.net

Like the triglycerides found in fats and oils (see page 28), these compounds are esters of glycerol and fatty acids. The compound on top is a **monoglyceride** as it has only **one** fatty acid linked to the glycerol, whereas

VIDEO LINK

Watch the clip on emulsifiers at www.brightredbooks.net

the compound below it has **two** fatty acids and is known as a **diglyceride**. It is the $-OH$ groups (coloured red) which make the heads of these molecules hydrophilic and therefore soluble in water. The tails, made up of non-polar fatty acid chains, are hydrophobic and are soluble in oil. These monoglycerides and diglycerides stabilise the emulsions and prevent the oil and water separating out. It should be noted that triglycerides cannot act as emulsifiers as they do not contain any $-OH$ groups to allow them to dissolve in water.

Examples of emulsions include ice cream, milk, salad cream, mayonnaise, butter, low-fat spreads and moisturising lotions.

DON'T FORGET

Monoglycerides and diglycerides, but not triglycerides, can be used as emulsifiers.

ONLINE TEST

Test yourself on this topic at www.brightredbooks.net

THINGS TO DO AND THINK ABOUT

The basic ingredients of mayonnaise are egg yolks, olive oil and vinegar. The egg yolks and vinegar are first beaten together and the oil is then added, drop by drop, with continual whisking until the emulsion is formed. With mayonnaise there is no need to add an artificial emulsifier because the egg yolk already contains a natural emulsifier. It is called lecithin. The branch (coloured red) has ionic charges and so is hydrophilic. The fatty acid branches on the left make up the hydrophobic part of the emulsifier.

FRAGRANCES

ESSENTIAL OILS

DON'T FORGET

Essential oils are the concentrated extracts of volatile compounds from plants that have pleasant aromas and are insoluble in water.

It has been known for thousands of years that many plants contain complex mixtures of volatile, pleasant-smelling substances that have come to be called **essential oils**. You just need to crush lavender leaves or peel an orange, for example, to release these volatile compounds into the air and experience their pleasant aromas. Essential oils can be extracted from plants by a process called 'steam distillation'. The water-insoluble oil that separates out usually has a smell characteristic of the particular plant from which it was extracted – for example, rose oil, geranium oil, clove oil, lavender oil, tea tree oil, eucalyptus oil and oil of turpentine.

Essential oils are widely used in perfumes, cosmetics, food flavourings, cleaning products, solvents and as alternative medicines – for example, in aromatherapy.

TERPENES

As organic chemistry advanced, chemists were able to separate the various components of essential oils, determine their molecular formulae and, later, their structural formulae. The most important and abundant constituents of essential oils are compounds called **terpenes** and more than 20 000 different terpenes are known. They may be unsaturated hydrocarbons, or they may contain oxygen and be alcohols, ketones or aldehydes. Oxygen-containing terpenes are normally referred to as **terpenoids**.

After analysing a large number of terpenes, it was realised that most have carbon skeletons containing 10, 15, 20, 25, 30 or 40 carbon atoms – that is, they contain carbon atoms in multiples of five. This suggested that there was a compound with five carbon atoms that served as a building block for the synthesis of terpenes. The building block was identified as **isoprene**, which has the molecular formula C_5H_8 and the structural formula:

DON'T FORGET

You must be able to draw a structural formula for the isoprene unit and name it as 2-methylbuta-1,3-diene.

The systematic name of isoprene is **2-methylbuta-1,3-diene**.

ONLINE

Read more about terpenes at www.brightredbooks.net

One of the simplest terpenes is **myrcene**, which is found in bay leaves. It has the molecular formula $C_{10}H_{16}$ and the structural formula shown on the right.

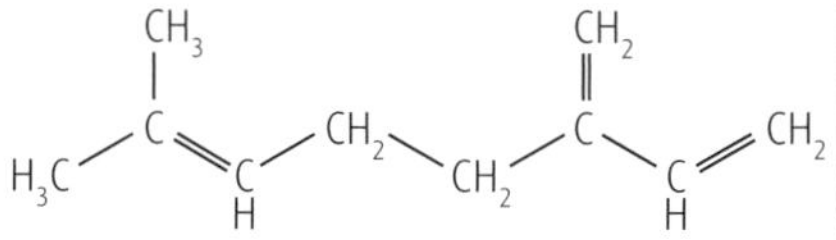

As it contains ten carbon atoms, it must be made by linking together two C_5 isoprene units:

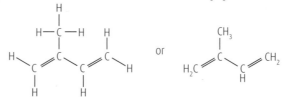

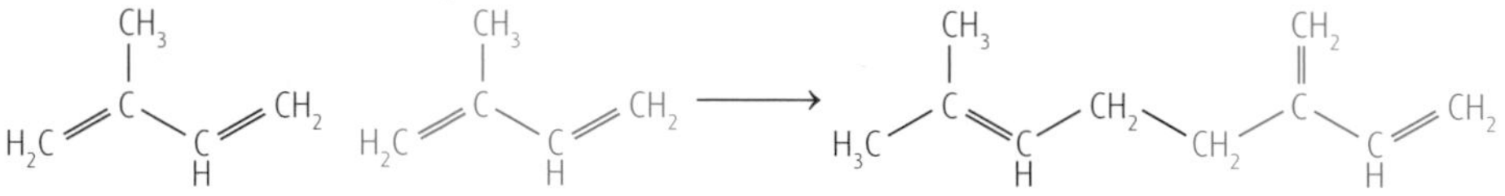

myrcene

Limonene ($C_{10}H_{16}$), a terpene found in oranges and lemons, is an isomer of myrcene, but has a ring structure. The two isoprene units link together as shown on the right.

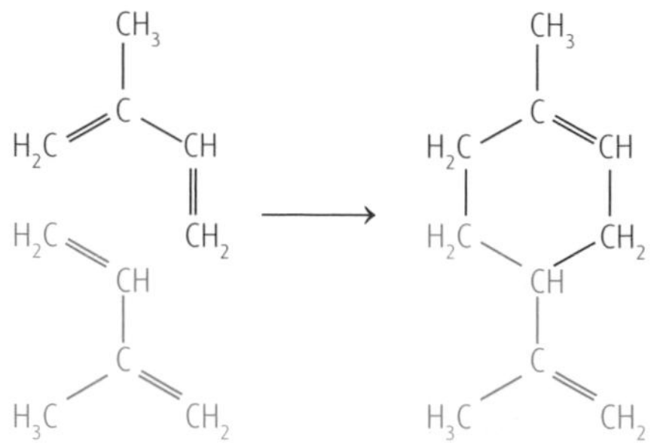

Some terpenes can be oxidised in the plant to give the so-called **terpenoids** (oxygen-containing terpenes). The structures of four of these terpenoids are drawn on the top right of page 43 with the isoprene units shown in colour.

limonene

contd

You will notice from the name endings that menth**ol** and farnes**ol** are alcohols, citronell**al** is an aldehyde and carv**one** is a ketone. Also notice that citronellal, carvone and menthol are C_{10} terpenoids (two isoprene units linked together), whereas farnesol is a C_{15} terpenoid (three isoprene units linked together).

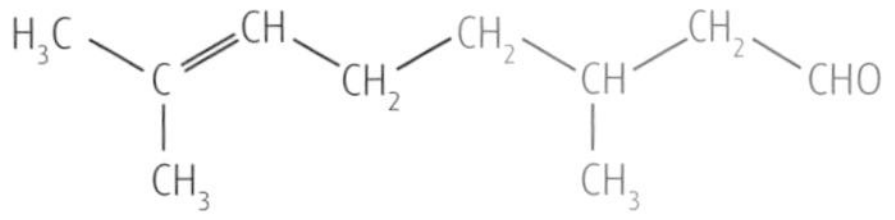

citronellal
(lemon oil)

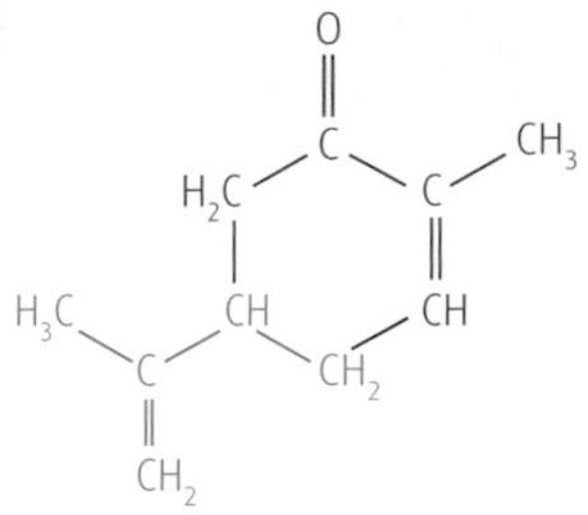

carvone
(spearmint oil)

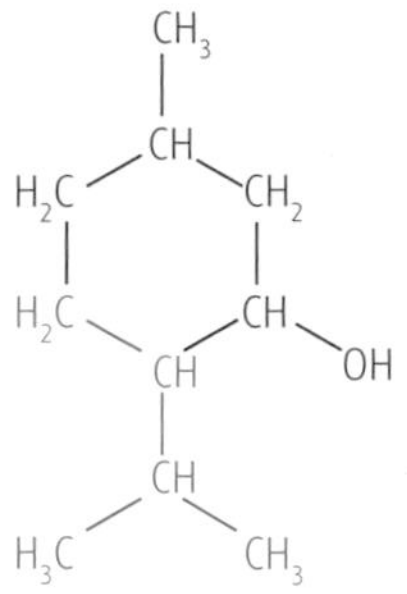

menthol
(peppermint oil)

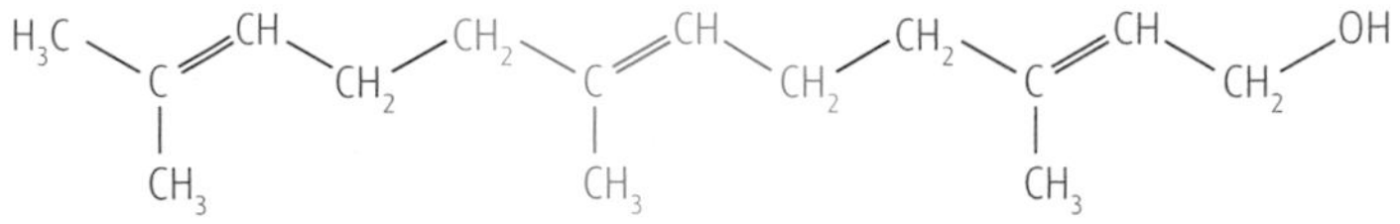

farnesol
(rose oil)

ESSENTIAL OILS, EDIBLE OILS AND MINERAL OILS

The use of the term **oil** has led to a lot of confusion between essential oils, edible oils and mineral oils and it is important that we are clear about the differences between these three types of oil.

As we have just found out, **essential oils** are the concentrated extracts of the volatile aroma compounds from plants and the key components of essential oils are **terpenes** and **terpenoids**.

Edible oils are also found in plants, but these are made up of triglycerides – that is, **esters** of the alcohol glycerol and the carboxylic acids known as fatty acids.

Mineral oils, on the other hand, are derived from crude oil and contain a variety of long-chain **alkanes** in the range C_{15} to C_{40}.

DON'T FORGET

You can work out the number of isoprene units (C_5 units) linked together in a terpene or terpenoid by dividing the total number of carbon atoms in its structure by five.

THINGS TO DO AND THINK ABOUT

1 A structural formula for isoprene is drawn below with its skeletal formula alongside:

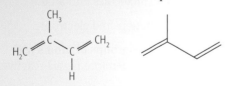

Draw skeletal formulae for myrcene and limonene. You will find structural formulae for myrcene and limonene on the previous page.

2 When the bark of a rubber tree is cut, or the stalks of dandelions are broken, a sticky white liquid, known as latex, oozes out. The latex contains **natural rubber**, a polymer which, like terpenes, is built up from isoprene units. On average, a molecule of rubber contains 5000 isoprene units. The structure of part of this polymer chain is drawn below with the isoprene units highlighted

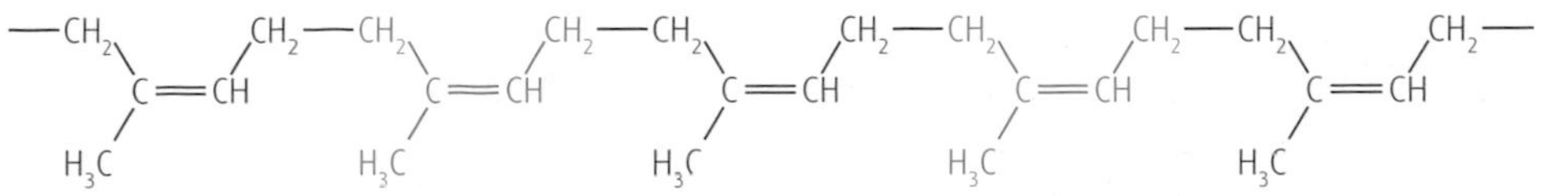

Natural rubber may be regarded as a polyterpene.

ONLINE TEST

Test your knowledge of fragrances online at www.brightredbooks.net

SKIN CARE

EFFECT OF ULTRAVIOLET LIGHT

Ultraviolet (UV) light makes up part of sunlight and, because it is a high-energy form of radiation, it can have damaging effects on our skin. The reason for this is that the energy associated with UV radiation is sufficient to break chemical bonds in molecules, leading to reactions which destroy skin tissue.

Too much exposure to the sun's UV radiation causes **sunburn** and the results – redness, blistered and peeling skin, and pain – are the same as the burns produced by intense heat. Repeated exposure to UV radiation over a longer period may produce **skin cancers**, especially among fair-haired, light-skinned people. Here, the UV radiation breaks bonds in DNA molecules, causing damage to genes. On a slightly less serious level, exposure to the sun over years leads to **aging of the skin**, giving it a wrinkly and leathery appearance. This is caused by the UV radiation destroying collagen, one of the proteins responsible for the firmness of skin.

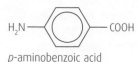

p-aminobenzoic acid

The most effective way of protecting skin from the damaging effects of UV radiation is to use **sun-block** products. They contain zinc oxide and titanium dioxide, which reflect UV radiation and do not allow any UV radiation through to the skin. **Sun-screen** products can also be used, but are less effective. They contain chemicals, such as *p*-aminobenzoic acid, shown above which absorbs some of the UV radiation so less reaches the skin.

On a more positive note, one beneficial effect of absorbing small amounts of UV radiation into the skin is the generation of **vitamin D**.

FREE RADICAL REACTIONS

You may be already familiar with the rapid addition reaction that takes place when bromine is added to an alkene. Alkanes can also react with bromine, but only in the presence of sunlight and, in particular, UV light. For example:

$$CH_4 + Br_2 \rightarrow CH_3Br + HBr$$

The reaction is thought to proceed by way of a **chain reaction**, which is characterised by three main stages: **initiation**, **propagation** and **termination**.

Initiation step

UV light plays a major part in the initiation step. It provides the energy required to break the bromine–bromine bond and to split some of the bromine molecules into atoms:

$Br{-}Br \rightarrow Br\bullet + Br\bullet$ (the bold dot in Br• represents an unpaired electron)

Bromine atoms are examples of **free radicals** – that is, **atoms or groups of atoms with an unpaired electron**. Free radicals are extremely unstable and therefore highly reactive.

The initiation step in a chain reaction is the process whereby free radicals are generated. You may be wondering why it was the Br–Br bond that was broken initially and not a C–H bond in the methane. The reason for this is simply that a Br–Br bond is weaker than a C–H bond. The Br–Br bond enthalpy is 194 kJ mol^{-1} and the C–H bond enthalpy is 412 kJ mol^{-1}.

DON'T FORGET

A free radical is an atom or group of atoms with an unpaired electron and is highly reactive.

Propagation steps

Each bromine free radical that is produced in the initiation step goes on to attack a methane molecule, removing a hydrogen atom from it to form hydrogen bromide and a methyl free radical. The latter then attacks a bromine molecule to form bromomethane and another bromine free radical as shown on the right.

(i) $Br\bullet + CH_4 \rightarrow HBr + CH_3\bullet$
(ii) $CH_3\bullet + Br_2 \rightarrow CH_3Br + Br\bullet$

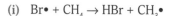

contd

In each of the propagation steps, one free radical enters the reaction and another free radical is generated. These steps 'propagate' or sustain the chain reaction.

Termination steps

As the number of free radicals builds up, collisions between them occur and stable molecules are produced.

(i) $CH_3\bullet + Br\bullet \rightarrow CH_3Br$
(ii) $CH_3\bullet + CH_3\bullet \rightarrow CH_3CH_3$
(iii) $Br\bullet + Br\bullet \rightarrow Br_2$

The termination steps are reactions in which free radicals are used up and not regenerated. Such reactions bring the chain reaction to an end.

Another example of a free radical chain reaction initiated by UV light is that between hydrogen and chlorine to form hydrogen chloride:

Initiation: $\quad Cl_2 \rightarrow 2Cl\bullet$
Propagation: $\quad Cl\bullet + H_2 \rightarrow HCl + H\bullet$ and $H\bullet + Cl_2 \rightarrow HCl + Cl\bullet$
Termination: $\quad H\bullet + Cl\bullet \rightarrow HCl$ and $Cl\bullet + Cl\bullet \rightarrow Cl_2$ and $H\bullet + H\bullet \rightarrow H_2$

In this case, the chain reaction is so rapid that it results in an explosion.

FREE RADICAL SCAVENGERS

When UV radiation interacts with compounds in the skin, free radicals are produced. In turn, these free radicals can initiate chain reactions which can cause severe damage to the cells. To counteract this, compounds known as **free radical scavengers** are used. **They react with free radicals to form stable molecules**, thus stopping the chain reactions from taking place. Vitamins C and E are common free radical scavengers added to skin creams and cosmetics to reduce the damaging effects caused by exposure to UV radiation.

The oxidation reactions that occur when food 'spoils' also involve the production of free radicals and this explains why free radical scavengers such as vitamins C and E and butylated hydroxyanisole (BHA) and butylated hydroxytoluene (BHT) are added to food products. BHA and BHT are often incorporated into plastics to prevent their disintegration on prolonged exposure to UV light from the sun.

The structures of vitamin C, vitamin E, BHA and BHT can be found on page 39.

 DON'T FORGET

A free radical chain reaction is characterised by three steps: initiation, propagation and termination.

 VIDEO LINK

For a demonstration of the hydrogen/chlorine explosive reaction, visit www.brightredbooks.net

DON'T FORGET

A free radical scavenger is a molecule which can react with free radicals to form stable molecules and prevent chain reactions.

 ONLINE TEST

Test yourself on skin care online at www.brightredbooks.net

 ## THINGS TO DO AND THINK ABOUT

1 Fluorine reacts with methane through a free radical chain reaction. Some steps in the chain reaction are shown in the table on the right.

   a Name the first step.
   b Write an equation for the other propagation step.
   c Write equations for the other two termination steps.

Reaction step	Name of step
$F_2 \rightarrow 2F\cdot$	
$F\cdot + CH_4 \rightarrow HF + CH_3\cdot$	Propagation
$CH_3\cdot + F\cdot \rightarrow CH_3F$	Termination

2 The antioxidant, compound **A**, can prevent damage to the skin by reacting with free radicals such as $NO_2\bullet$.

Why can compound **A** be described as a free radical scavenger in this reaction?

compound **A**

CHEMISTRY IN SOCIETY

GETTING THE MOST FROM REACTANTS 1

THE UK CHEMICAL INDUSTRY

The UK chemical industry is a major contributor to the quality of our everyday lives. Without this chemical industry, thousands of consumer products that we currently take for granted would not exist. For example, there would be no plastics, synthetic fibres, detergents, pharmaceuticals, fuels such as petrol and diesel, paints, toiletries, and so on. The chemical industry also has an important role in our national economy. It is the UK's largest manufacturing export sector, contributing £30 million every day to the country's balance of trade with the rest of the world. It also provides direct employment for about 250 000 people and indirectly supports several hundred thousand additional jobs.

Industrial chemical processes are designed to make as much profit as possible for investors, while minimising any negative impacts on the environment.

FACTORS INFLUENCING PROCESS DESIGN

Availability, sustainability and cost of feedstocks

Feedstocks are the reactants which go into a chemical process. These feedstocks are usually extracted from raw materials and then purified before use. If the raw materials are widely available and the feedstocks are easily extracted and purified, then the cost of the feedstocks will be low. If, however, the raw materials are imported and the feedstocks are difficult to extract, then the process is much more expensive. Two typical raw materials that are both cheap and widely available are water and air. Other useful raw materials include fossil fuels, minerals and metal ores. Some of these are found in the UK, but many need to be imported. The cost of these raw materials varies depending on currency exchange rates and the economic and political stability of the country from which they are imported. A major concern in the twenty-first century is that many raw materials, such as crude oil, are running out. Resources need to be sustainable: a sustainable resource is one that can be renewed at the same speed or faster than it is used up. In the manufacture of paper, for example, wood is a sustainable resource as long as the trees are harvested no faster than new trees can mature. This requires the careful management of forests and a programme of planting new trees. Wind, solar and hydroelectric power can be considered as sustainable sources of energy, whereas oil, natural gas and minerals taken from the Earth do not regenerate and are non-sustainable.

Energy requirements

The efficient use of energy is an important consideration in most chemical processes because of its high cost. Many chemical reactions are exothermic and the heat energy released can be conserved by lagging pipes and using heat exchangers. Energy from the exothermic steps in a process can be used to supply energy for the endothermic steps. Using catalysts which operate at lower temperatures is another way of reducing energy costs.

Opportunities for recycling

In an effort to reduce the natural resources used and to minimise waste, every opportunity should be taken by the chemical industry to recycle materials. This is also better for the environment and should make processes cheaper because fewer raw materials need to be purchased.

DON'T FORGET

Important factors which influence process design include:
- the availability, sustainability and cost of raw materials and feedstocks
- energy requirements
- recycling opportunities
- the marketability of by-products
- the yield of products.

contd

Marketability of by-products

In many chemical processes a by-product is formed in addition to the desired product. If this by-product cannot be used, then this is a costly waste of resources. However, if the by-product is useful and can be sold, then this may make the difference between a process being ruled out because it is too expensive and the process becoming profitable.

Product yield

The chemical industry exists to create wealth and wealth is only created if the industry makes a profit. It is therefore important that the yield of products is as high as possible. This will be dealt with later, but chemists try to find the combination of temperature, pressure and concentrations of reactants that will give the highest yield of product, preferably at the lowest cost.

CHOICE OF SYNTHETIC ROUTE

A chemical can usually be made in a number of different ways. Many factors influence which synthetic route is chosen to manufacture a particular chemical. Ethanoic acid, for example, can be made either by the direct oxidation of naphtha or by the catalysed reaction between methanol and carbon monoxide. In both reactions, the operating conditions of temperature and pressure are roughly the same. The first process has the advantage that no expensive catalyst is needed, but the yield of ethanoic acid is very low. This low yield implies that considerable amounts of by-products are formed, but, fortunately, buyers can be found for these substances so the manufacture of ethanoic acid from naphtha returns a healthy profit. The major advantage of the second process is the high yield of ethanoic acid (over 99%) and the flexibility of feedstock production – methanol can be made from any fossil fuel. The major disadvantage is the high cost of the catalyst. The methanol/carbon monoxide route to ethanoic acid is now preferred and all new plants use this process. Existing naphtha oxidation plants are unlikely to be replaced, however, until the maintenance costs become too high or there is no market for the by-products.

ENVIRONMENTAL CONSIDERATIONS

These are major considerations in the chemical industry and are subject to rigorous legislation. Computer systems, for example, are used to monitor the process and will automatically shut down any part of the plant which is not functioning correctly. The air quality in the vicinity of chemical plants is also monitored and there are warning systems to alert local residents if any abnormal emissions are detected.

Modern chemical plants are designed to minimise waste and to avoid, where possible, the use and production of toxic substances. Where appropriate, it is better for the environment to make products that are biodegradable.

 ## THINGS TO DO AND THINK ABOUT

The term 'chemical' conjures up 'toxic substance' in the minds of many members of the general public. They do not appreciate that everything in the world, including human beings, is made up of chemicals, most of which are harmless. Any accident in the chemical industry attracts bad publicity, fuelling the public's negative image of the industry. Although the only acceptable accident rate is zero, remember that working in the chemical industry is significantly less hazardous than crossing the street. Although the industry itself tries hard to improve its image, we, as chemists, must help with this process.

 DON'T FORGET

The choice of a particular synthetic route is influenced by factors such as the cost, availability and suitability of feedstocks, the product yield, opportunities for the recycling of reactants and the marketability of any by-products.

 DON'T FORGET

Chemists design new molecules and new materials. 'Green' chemists make sure that their products not only do what they are supposed to do, but that they do it with the least cost to the environment.

 **ONLINE**

Much of what has been mentioned in these pages now comes under the term 'green chemistry'. It is often stated that there are 12 principles of green chemistry. You will find much more about this at www.brightredbooks.net

 ONLINE TEST

Take the 'Getting the most from reactants' test at www.brightredbooks.net

 DON'T FORGET

Most of the materials in everyday use which we take for granted would not be present if it was not for the chemical industry.

GETTING THE MOST FROM REACTANTS 2

THE MOLE

The **mole** is a central concept in chemistry. One mole of a substance can be considered as its gram formula mass. We can think of the mole as the unit of chemical currency in balanced chemical equations and calculations.

For example, we can write the balanced chemical equation for aluminium burning in oxygen to produce aluminium oxide as:
$4Al + 3O_2 \rightarrow 2Al_2O_3$

At a molecular level, we may consider this as four aluminium atoms reacting with three oxygen molecules.

However, if we rewrite the equation as:
$2Al + 1\frac{1}{2}O_2 \rightarrow Al_2O_3$

then it does not make sense to think of two aluminium atoms reacting with one and a half oxygen molecules.

What balanced equations show is the mole ratio(s) of the reactants and products. Look at the first equation again:
$4Al + 3O_2 \rightarrow 2Al_2O_3$

This shows that four **moles** of aluminium atoms react with three **moles** of oxygen molecules to produce two **moles** of aluminium oxide.

So, when written as:
$2Al + 1\frac{1}{2}O_2 \rightarrow Al_2O_3$

it shows that two **moles** of aluminium atoms react with one and a half **moles** of oxygen molecules to produce one **mole** of aluminium oxide.

NOTE

The word 'mole' is often abbreviated to 'mol'.

You should try to get used to writing balanced chemical equations with 'half' values in them. This will be important later on when writing balanced equations for enthalpy changes.

If we put values into the equation:

$2Al$	$+ 1\frac{1}{2}O_2$	$\rightarrow$	Al_2O_3
2 mol	$+ 1\frac{1}{2}$mol	$\rightarrow$	1 mol
(2 × 27 g)	+ ($1\frac{1}{2}$ × 32 g)	$\rightarrow$	1 × 102 g
54 g	+ 48 g	$\rightarrow$	102 g

This shows that no mass is gained or lost in a chemical reaction. The total mass of the products is always equal to the total mass of the reactants.

DON'T FORGET

Given the mass of a substance, the formula to use is $n = \frac{m}{GFM}$, which can be found on page 4 of the Data Booklet.

MASS TO MASS CALCULATIONS

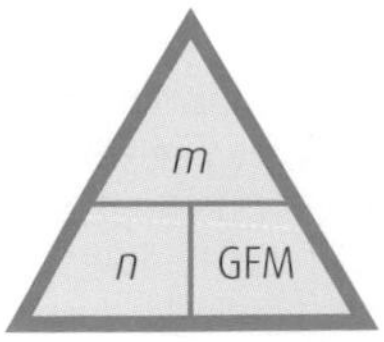

As a mole is the gram formula mass and is a measurable quantity, we can use balanced equations to calculate the mass of a reactant or product in a chemical reaction.

To calculate the number of moles of a substance from the mass of the substance, we can use the relationship:
number of moles, $n = \frac{\text{mass}}{\text{gram formula mass}}$ or $n = \frac{m}{GFM}$

Some students prefer to use a formula triangle as shown on the left to help remember this relationship.

Example: Sample calculation

What mass of aluminium oxide would be formed if 6·0 g of aluminium is burned in excess oxygen? Note that the oxygen must be in excess so that all the aluminium is completely reacted.

contd

Answer:

First calculate the number of moles of aluminium:

$n = \frac{\text{mass}}{\text{gram formula mass}}$ or $n = \frac{m}{\text{GFM}} = \frac{6 \cdot 0}{27 \cdot 0} = 0 \cdot 22$ mol

The balanced equation is $2Al + 1\frac{1}{2} O_2 \rightarrow Al_2O_3$ and from this we can see that 2 mol of Al will give 1 mol of Al_2O_3.

Therefore 0·22 mol of Al will give 0·11 mol of Al_2O_3.

The mass of 0·11 mol of $Al_2O_3 = n \times \text{GFM} = 0 \cdot 11 \times 102 = $ **11·22** g.

MOLAR VOLUMES OF GASES

The mass of one mole of different substances is likely to be different. However, although the mass may be different, the volume of one mole of different gases is the same under the same conditions of temperature and pressure.

For example, this means that one mole of hydrogen weighing 2·0 g will occupy the same volume as one mole of sulfur dioxide weighing 64·1 g at the same temperature and pressure.

It is also true to state that 0·20 g of hydrogen (0·1 mol) would have the same volume as 6·41 g of sulfur dioxide (0·1 mol) at the same temperature and pressure.

The volume of one mole of gas is known as the molar volume.

At room temperature, which is about 25°C, and one atmosphere pressure, the molar volume of any gas is about 24 litres per mole.

 **DON'T FORGET**

The molar volume is the same for all **gases** at the **same temperature and pressure**.

SIMPLE CALCULATIONS USING MOLAR VOLUMES

The fact that all gases have the same molar volume allows us to calculate the volume of a gas from the number of moles and vice versa.

Example:

Calculate:

(a) The number of moles in 0·60 litres of oxygen.

(b) The mass of 0·60 litres of oxygen.

(Take the volume of one mole of oxygen to be 24.0 litres.)

Answer:

(a) Molar volume = 24·0 l mol^{-1}. Therefore the number of moles is equal to 0·60 l divided by the molar volume:
$n = \frac{0 \cdot 60}{24 \cdot 0} = 0 \cdot 025$ mol.

(b) Mass = $n \times \text{GFM} = 0 \cdot 025 \times 32 \cdot 0 = 0 \cdot 8$ g.

  THINGS TO DO AND THINK ABOUT

1 Calculate
   a the mass of magnesium oxide produced when 4·86 g of magnesium burns in an excess of oxygen.
   b the mass of calcium oxide formed when 2·05 g of calcium carbonate are heated to produce calcium oxide and carbon dioxide.

2 Which of the following quantities of gases occupies the largest volume under the same conditions of temperature and pressure?
   A 0·2 g of hydrogen
   B 3·2 g of oxygen
   C 0·8 g of helium
   D 2·8 g of nitrogen.

3 The mole is not just the gram formula mass of a substance. It can also be expressed as a number. One mole of any substance contains $6 \cdot 02 \times 10^{23}$ formula units. This number is known as the Avogadro constant and has the symbol L. Calculate the number of water molecules in 180 g of water given that the formula units for water are H_2O molecules.

 ONLINE TEST

Take the 'Getting the most from reactants' test at www.brightredbooks.net

GETTING THE MOST FROM REACTANTS 4

PERCENTAGE YIELDS

Types of yield

The yield of a chemical reaction is the amount of product obtained in the reaction. There are two types of yield:

- The theoretical yield - this is the maximum amount of product that could be obtained if there was 100% conversion of reactants into products.
- The actual yield – this is the amount of product that is obtained in practice.

The actual yield is usually less than the theoretical yield. There can be several reasons for this, including:

- The reaction may be reversible, which means that the conversion of reactants into products will never be 100%.
- Side reactions may occur in addition to the main reaction; the formation of the side products will inevitably reduce the yield of the main product.
- Even when 100% conversion is achieved, some of the product is likely to be lost when it is separated from the reaction mixture and purified.
- The initial reactants may not be 100% pure.

The actual yield of product can be expressed as a percentage of the theoretical yield:

$$\text{Percentage yield} = \frac{\text{actual yield}}{\text{theoretical yield}} \times 100$$

Percentage yield calculations

Example: 1

When excess ethanoic anhydride was added to 14·4 g of salicylic acid, 6·26 g of aspirin were obtained. Calculate the percentage yield of aspirin given that the balanced equation for the reaction is:

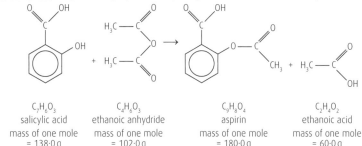

$C_7H_6O_3$	$C_4H_6O_3$	$C_9H_8O_4$	$C_2H_4O_2$
salicylic acid	ethanoic anhydride	aspirin	ethanoic acid
mass of one mole = 138·0 g	mass of one mole = 102·0 g	mass of one mole = 180·0 g	mass of one mole = 60·0 g

Answer:

We must first calculate the theoretical yield of aspirin assuming a 100% conversion of reactants into products. We can do this in the usual way using the balanced equation for the reaction.
Although the mass of one mole of all the reactants and products are given in the question, we really only need the mass of one mole of salicylic acid and aspirin in the calculation.

$$C_7H_6O_3 \ + \ C_4H_6O_3 \rightarrow C_9H_8O_4 + C_2H_4O_2$$

1 mol	1 mol
138·0 g	180·0 g
14·4 g	$180\cdot0 \text{ g} \times \frac{14\cdot4}{138\cdot0} = 18\cdot78$ g

The theoretical yield of aspirin is 18·78 g and the actual yield is 6·26 g. So by substituting these values into the percentage yield expression we obtain:

$$\text{Percentage yield} = \frac{\text{actual yield}}{\text{theoretical yield}} \times 100 = \frac{6\cdot26}{18\cdot78} \times 100 = \mathbf{33\cdot3\%}$$

contd

Example: 2

Ammonia is made by the Haber Process:

$N_2(g) + 3H_2(g) \rightarrow 2NH_3(g)$

Under test conditions, the percentage yield of ammonia in the Haber Process was 14%. Calculate the mass of hydrogen that would be needed to react with excess nitrogen to give an actual yield of 150 kg of ammonia.

Answer:

We have been given the percentage yield and the actual yield of ammonia and so, by rearranging the percentage yield expression, we can work out the theoretical yield of ammonia – that is, the yield for 100% conversion.

$$\text{percentage yield} = \frac{\text{actual yield}}{\text{theoretical yield}} \times 100$$

Rearrange to:

$$\text{theoretical yield} = \frac{\text{actual yield}}{\text{percentage yield}} \times 100$$

$$\text{theoretical yield} = \frac{150}{14} \times 100 = 1071 \text{ kg}$$

Now we can use the balanced equation to calculate the mass of hydrogen that would be needed:

$$N_2 + 3H_2 \rightarrow 2NH_3$$
$$3 \text{ mol} \leftrightarrow 2 \text{ mol}$$
$$6{\cdot}0 \text{ g} \leftrightarrow 34{\cdot}0 \text{ g}$$
$$6{\cdot}0 \times \frac{1071 \times 10^3}{34{\cdot}0} \leftrightarrow 1071 \times 10^3 \text{ g}$$
$$= 189 \times 10^3 \text{ g}$$
$$= \mathbf{189 \text{ kg}}$$

VIDEO LINK

Learn more about percentage yield calculations by watching the clip at www.brightredbooks.net

 ## THINGS TO DO AND THINK ABOUT

ONLINE TEST

Take the 'Getting the most from reactants' test at www.brightredbooks.net

Many substances are made, not by a single reaction step, but by a sequence of steps. So, how will the overall percentage yield of the desired product be affected by the percentage yields in the individual steps? To help answer this question, suppose compound C was made in two steps, starting with compound A:

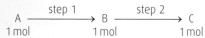

$$A \xrightarrow{\text{step 1}} B \xrightarrow{\text{step 2}} C$$
$$1 \text{ mol} \qquad 1 \text{ mol} \qquad 1 \text{ mol}$$

Let us further suppose that the percentage yield in step 1 was 70% and that the percentage yield in step 2 was 60%.

It is tempting to say that the overall percentage yield of C would be an average of the two – that is, 65% – but this would be naive. If we started with 1·00 mol of A, then the yield of B would be 0·70 mol as the percentage yield in step 1 is 70%. In step 2, only 60% of the 0·70 mol of B would be converted into C. So the actual yield of C would be 0·42 mol compared with a theoretical yield of 1·00 mol.

This implies that the overall percentage yield $= \frac{0{\cdot}42}{1{\cdot}00} \times 100 = 42\%$

This represents a dramatic fall from the 70% and 60% yields in the individual steps. A situation like this would not be tolerated in the chemical industry and explains why chemists are always seeking to find direct routes with high percentage yields when synthesising chemicals.

GETTING THE MOST FROM REACTANTS 5

ATOM ECONOMY

Percentage yield and atom economy

You now know how to calculate percentage yield using the relationship:

$$\% \text{ yield} = \frac{\text{actual yield}}{\text{theoretical yield}} \times 100$$

The percentage yield measures how effectively a particular industrial process converts expensive reagents into the desired final product. Chemists are keen to reduce waste, both to help protect the environment and to keep production costs to a minimum. To compare how well different methods avoid potential waste, chemists often calculate the **atom economy** for each of the possible reactions. Atom economy is defined as:

$$\text{atom economy} = \frac{\text{mass of desired product in equation}}{\text{total mass of reactants in equation}} \times 100$$

The atom economy measures the proportion of the total mass of all starting materials successfully converted into the desired product. The higher the atom economy figure, the less waste is likely to be produced.

Reactions which have a high percentage yield may have a low atom economy value if large amounts of unwanted by-products are formed.

A major reason for the shift to atom economy in place of percentage yield is that the costs of treating unwanted waste products from industrial chemical reactions means that the chemical industry must now consider the economic benefits of applying the principles of green chemistry (see page 47).

Green chemistry encourages environmentally conscious behaviour, such as reducing and preventing pollution and the destruction of our planet.

The atom economy of a chemical reaction depends on the type of reaction involved. For example, if only one product is formed in a reaction, the atom economy for that reaction will be 100% as there will be no waste or by-products formed.

Calculating atom economy

Example: 1

1,2-Dibromopropane can be manufactured by reacting propene with bromine:

C_3H_6	+	Br_2	$\rightarrow$	$C_3H_6Br_2$
1 mol		1 mol		1 mol
42·0 g		159·8 g		201·8 g

Total mass of reactants = 201·8 g

Mass of desired product (1,2-dibromopropane) = 201·8 g

$$\text{atom economy} = \frac{\text{mass of desired product in equation}}{\text{total mass of reactants in equation}} \times 100 = \frac{201·8}{201·8} \times 100 = 100\%$$

In terms of atom economy, this reaction is 100% efficient as all the reactant atoms are included within the desired product.

contd

DON'T FORGET

The atom economy will always be 100% in reactions in which only one product is formed.

Example: 2

Nitrogen monoxide, NO, an important gas in the manufacture of nitric acid, can be made by reacting ammonia, NH_3, with oxygen:

$$4NH_3(g) + 5O_2(g) \rightarrow 4NO(g) + 6H_2O(g)$$

4 mol   5 mol   4 mol   6 mol

68·0 g   + 160·0 g $\rightarrow$ 120·0 g + 108·0 g

Total mass of reactants = 228·0 g

Mass of desired product (nitrogen monoxide) = 120·0 g

atom economy $= \frac{\text{mass of desired product in equation}}{\text{total mass of reactants in equation}} \times 100 = \frac{120 \cdot 0}{228 \cdot 0} \times 100 = 52 \cdot 6\%$.

Therefore this reaction has an atom economy of less than 53% as many of the reactant atoms end up in water molecules rather than in the desired product.

Example: 3

On page 53 you were told in Example 2 that the percentage yield for the production of ammonia in the Haber Process was 14%.

Calculate the atom economy in the Haber Process.

The balanced equation in the Haber Process is $N_2 + H_2 \rightarrow 2NH_3$.

As all the reactant atoms change into the product and there are no waste or by-products, then the atom economy is 100%.

ONLINE

Find more information on atom economy by following the link at www.brightredbooks.net which includes a PowerPoint presentation with more atom economy calculations.

THINGS TO DO AND THINK ABOUT

Manufacture of ibuprofen

Ibuprofen was first patented and manufactured by the Boots Company in the 1960s. The Boots' synthesis started with a hydrocarbon known as 2-methylpropylbenzene, which was made from compounds present in crude oil. The synthesis of ibuprofen from 2-methylpropylbenzene required six steps and so, even if each step had a 90% yield, then the overall percentage yield would be approximately 53%.

As the total gram formula mass of all the reactants used in the Boots' synthesis is 514·5 g and ibuprofen has the molecular formula $C_{13}H_{18}O_2$ with a gram formula mass of 206·0 **g**, then the atom economy is approximately 40%.

When Boots' patent for manufacturing ibuprofen expired in the 1980s, another company known as BHC developed a 'greener' synthesis of ibuprofen. Only three steps were involved and the percentage yield was higher than the Boots' synthesis. The BHC method also starts with 2-methylpropylbenzene, but this time the overall atom economy is 77% as fewer other reactants are used and there is less waste and by-products.

1 The BHC synthesis of ibuprofen has another advantage over the Boots' synthesis which makes it less damaging to the environment. Use the website links to find out what this is.

2 The percentage yield in the reaction forming ammonia in the Haber Process is very low (about 14%), but it has an atom economy of 100%. There are no by-products. Suggest a way that the percentage yield could be increased in the Haber Process.

ONLINE TEST

Take the 'Getting the most from reactants' test at www.brightredbooks.net

ONLINE

You will find much more information on the structure and synthesis of ibuprofen by following the links at www.brightredbooks.net, which include a case study in green chemistry and gives more information on how ibuprofen is currently prepared.

GETTING THE MOST FROM REACTANTS 6

THE IDEA OF EXCESS

The reactant in excess and the limiting reactant

In the chemical industry, when manufacturing a product, one particular reactant will often be significantly more expensive than the others. The chemist in charge of production will often deliberately use slightly more of the cheaper ingredients than the balanced equation suggests to try to make sure that all of the more valuable chemical reacts. In this way, even if there are impurities present in the cheaper feedstocks, there will still be plenty of these substances present to react with the expensive ingredient. Later you will learn how, for reversible reactions, using excess amounts of the less expensive ingredients also helps to increase the percentage yield for the reaction.

The reactant that there is too much of is said to be '**in excess**' and some of this reactant will be left over. Sometimes it is necessary to calculate which reactant is in excess. The reactant which is not in excess will all be used up and so it can be used to calculate how much product will be formed. This is known as the **limiting reactant** as it limits the amount of product that can be formed. The chemical reaction will stop when all the limiting reactant has been used up.

The idea of an excess reactant and a limiting reactant can sometimes be confusing. One way of looking at this is to consider the relationship between cars and tyres. There are four tyres per car and, in chemical reaction terms, we can think of this as the mole ratio.

If there are 4 cars and 16 tyres, then this is the correct ratio. However, if there are 4 cars and only 13 tyres, then we can say there is an excess of cars. If there are 4 cars and 20 tyres, then the tyres are in excess. No matter how many tyres there are in excess of 16, only 4 cars can be fitted with tyres if there are only 4 car bodies. Likewise, in a chemical reaction, if there is only a certain amount of one reactant available, then the reaction must stop when that limiting reactant is consumed, whether or not all of the other reactant has been used up.

To work out which reactant is in excess and which is the limiting reactant, it is necessary to calculate the number of moles of each reactant and to use these values in the balanced formula equation.

Which reactant is in excess and which is the limiting reactant?

Which reactant would be in excess if 0·972 g of magnesium was added to 50 cm³ of 0·10 mol l⁻¹ hydrochloric acid?

Number of moles of Mg, $n = \frac{mass}{GFM} = \frac{0.972}{24.3} = 0.040$ mol.

The number of moles of HCl, $n = cV = 0.10 \times 0.050 = 0.0050$ mol.

The balanced equation is:
$Mg(s) + 2HCl(aq) \rightarrow MgCl_2(aq) + H_2(g)$

This shows that 1 mol of Mg will react with 2 mol of HCl.
So 0·040 mol of Mg would react with 0·080 mol of HCl.
As there is only 0·0050 mol of HCl present and 0·080 mol is required, then there is not enough HCl.
The Mg is therefore in excess and the HCl is the limiting reactant.

Calculating how much product will be formed

The amount of product formed can be calculated using the balanced formula equation and the number of moles of the limiting reactant –that is, the reactant that is **not** in excess.

DON'T FORGET

The number of moles in a pure substance, $n = \frac{mass}{GFM}$, see page 48.

DON'T FORGET

The number of moles of solute in a solution, $n = cV$, see page 51.

contd

Example:

Calculate the mass of hydrogen produced when 0·972 g of magnesium is added to 50 cm³ of 0·10 mol l⁻¹ hydrochloric acid.

Answer:

The balanced formula equation for the reaction is given below and it shows that 1 mol of Mg will react with 2 mol of HCl. The number of moles of each reactant has been calculated earlier and showed that magnesium is in excess and that the limiting reactant is the HCl. This means that the number of moles of HCl must be used in the calculation. This is shown below:

Balanced equation: $Mg(s) + 2HCl(aq) \rightarrow MgCl_2(aq) + H_2(g)$
From the equation: 1 mol   2 mol   1 mol   1 mol
From the number of moles of HCl calculated above: 0·0050 mol →   0·0025 mol

The equation shows that 1 mol of H_2 will be formed when 2 mol of HCl reacts completely, so 0·0025 mol of H_2 will be formed when 0·0050 mol of HCl reacts completely.

Mass of hydrogen formed = n × GFM = 0·0025 × 2 = **0·0050 g**.

 VIDEO LINK

To learn more about calculations with excesses, head to www.brightredbooks.net and watch the clip.

Another calculation

Calculate the mass of carbon dioxide produced when 0·506 g of magnesium carbonate is added to 100 cm³ of 0·50 mol l⁻¹ nitric acid.

First calculate the number of moles of each reactant.

The **mass** of magnesium carbonate is given.	The concentration and volume of nitric acid are given.
The formula of magnesium carbonate is $MgCO_3$ and the gram formula mass works out at 84·3 g	The volume of nitric acid is 100 cm³ which is 0·100 litres.
So n for $MgCO_3 = \frac{0·506}{84·3} = 0·0060$ mol	So n for $HNO_3 = 0·50 × 0·100 = 0·050$ mol

Then write the balanced chemical equation for the reaction:

$MgCO_3(s) + 2HNO_3 (aq) \rightarrow Mg(NO_3)_2(aq) + H_2O(l) + CO_2(g)$
1 mol   2 mol

The balanced equation tells us that one mole of $MgCO_3$ reacts with two moles of HNO_3. Therefore 0·0060 mol of $MgCO_3$ would react with 0·012 mol (from 0·0060 × 2) of HNO_3.

As there is 0·050 mol of HNO_3 present and only 0·012 mol will react, then HNO_3 is in excess.

This means that the limiting reactant is $MgCO_3$ and that all the $MgCO_3$ will be used up in the reaction.

Balanced equation: $MgCO_3(s) + 2HNO_3(aq) \rightarrow Mg(NO_3)_2(aq) + H_2O(l) + CO_2(g)$
From the equation: 1 mol   2 mol   1 mol   1 mol   1 mol

Therefore, 0·0060 mol of $MgCO_3$ will react to produce 0·0060 mol of CO_2.

The gram formula mass of CO_2 is 44 g, therefore the mass of CO_2 formed = n × GFM = 0·0060 × 44 = 0·264 g.

 ## THINGS TO DO AND THINK ABOUT

1   Calculate the mass of copper(II) carbonate required to react completely with 100 cm³ of 0·20 mol l⁻¹ sulfuric acid.

2   In each of the following examples, identify the reactant in excess and the limiting reactant and then calculate the number of moles of the salt formed.
    a   2·015 g of MgO added to 100 cm³ of 0·101 mol l⁻¹ HCl.
    b   4·000 g of $MgCO_3$ added to 200 cm³ of 0·010 mol l⁻¹ HNO_3.
    c   2·616 g of Zn added to 250 cm³ of 0·210 mol l⁻¹ H_2SO_4.

✓ **ONLINE TEST**

Take the 'Getting the most from reactants' test at www.brightredbooks.net

EQUILIBRIA 1

THE CONCEPT OF DYNAMIC EQUILIBRIUM

Reversible reactions

In a chemical reaction, reactants change into products. Some, such as boiling an egg or wool growing on a sheep, for example, most definitely cannot be reversed.

However, some reactions, such as the conversion of hydrated cobalt(II) chloride into anhydrous cobalt(II) chloride with the addition of heat, are reversible reactions – that is, the reaction can go backwards or forwards. This is shown in the two equations below:

$$CoCl_2.6H_2O \xrightarrow{\text{heat}} CoCl_2 + 6H_2O$$
$$\text{pink} \qquad\qquad \text{blue}$$
$$CoCl_2 + 6H_2O \rightarrow CoCl_2.6H_2O$$
$$\text{blue} \qquad\qquad \text{pink}$$

When pink hydrated cobalt(II) chloride is heated, the water is driven off and blue anhydrous cobalt(II) chloride is the product. When water is added to the blue anhydrous cobalt(II) chloride, it changes back to pink hydrated cobalt(II) chloride.

Another example of a reversible reaction is the thermal decomposition of ammonium chloride. This is carried out by heating ammonium chloride in a test-tube to produce ammonia and hydrogen chloride. The ammonia and hydrogen chloride produced then recombine further up the test-tube, forming ammonium chloride again.

This is shown in the equations below:

$$NH_4Cl(s) \rightarrow NH_3(g) + HCl(g) \quad \text{and} \quad NH_3(g) + HCl(g) \rightarrow NH_4Cl(s)$$

In fact, most chemical reactions are reversible. In a reversible reaction, the forward and reverse reactions occur at the same time and the reaction mixture contains both reactants and products.

Dynamic equilibrium

Consider the reversible reaction:

$$A + B \rightleftharpoons C + D$$

At the beginning of the reaction, when reactants A and B are at their most concentrated, the rate of the forward reaction will be at its greatest. As the reaction proceeds, A and B will be used up. This means that the concentrations of A and B are decreasing and therefore the rate of the forward reaction also decreases.

The products C and D are not present at the start and so, at the beginning, the rate of the reverse reaction will be zero. As the forward reaction proceeds, the products C and D will be formed and their concentrations gradually increase. The rate of the reverse reaction will therefore increase. As the rate of the forward reaction decreases and the rate of the reverse reaction increases, a balance point is eventually reached when the reaction appears to have stopped.

In fact, both the forward and the reverse reactions are still taking place, but at the same speed. We say that the reactions at this point have reached equilibrium. Reversible reactions attain a state of dynamic equilibrium when the rates of the forward and reverse reactions are equal.

The sign $\rightleftharpoons$ is used to show that a reaction is at dynamic equilibrium.

The changes in the rates of the forward and reverse reactions as equilibrium is being reached are summarised below:

contd

A + B ⇌ C + D   Forward reaction much faster than reverse reaction

A + B ⇌ C + D   Forward reaction slowing down, reverse reaction getting faster

A + B ⇌ C + D   Reverse reaction almost as fast as forward reaction

A + B ⇌ C + D   Both reactions occurring at the same rate, dynamic equilibrium has been reached

If the equilibrium concentrations of A and B are less than those of C and D, then the equilibrium position is said to lie to the right – that is, to the side of the products.

However, if the equilibrium concentrations of A and B are greater than those of C and D, then the equilibrium position lies to the left – that is, to the side of the reactants.

As the rate of the forward reaction is equal to the rate of the reverse reaction, then, at equilibrium, the concentrations of the reactants and products are constant (no longer changing). Note that this does not mean that the concentrations of the reactants and products are equal.

Position of equilibrium

Although the rates of the forward and reverse reactions are the same when equilibrium is reached, this does not mean that the concentrations of the reactants and products are the same. Once a reaction system is at equilibrium, it is impossible to tell whether the equilibrium mixture was obtained by starting with the reactants or starting with the products because the position of equilibrium is the same no matter which direction it is approached from.

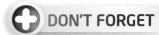

THINGS TO DO AND THINK ABOUT

Iodine dissolves in potassium iodide solution to produce a brown solution. Iodine dissolves in an organic solvent to produce a purple solution.

If these two iodine solutions are added together then, as the two solvents do not mix, an equilibrium is set up between the iodine dissolved in each of the solvents.

Test-tube A contains iodine dissolved in potassium iodide solution (brown) to which a colourless organic solvent is added. Iodine will start to move from the upper layer to the lower layer.

Test-tube B contains iodine dissolved in an organic solvent (purple) to which colourless potassium iodide solution is added. Iodine will start to move from the lower layer to the upper layer.

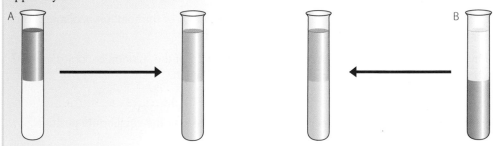

After some time the contents of both test-tubes look identical. Both have a brown solution sitting on top of a purple solution.

Why has this happened?

A position of equilibrium has been reached. The rate at which the iodine is moving from the potassium iodide solution downwards to the organic solvent is the same rate at which it is moving upwards from the organic solvent to the potassium iodide solution. This experiment illustrates that the position of equilibrium is the same, no matter from which direction it is approached.

EQUILIBRIA 2

SHIFTING THE EQUILIBRIUM POSITION: THE EFFECT OF CHANGING CONCENTRATION AND PRESSURE

Many chemical reactions are reversible and so the products may be in equilibrium with the reactants. This may result in costly reactants not being completely converted into the required products. In a closed system, dynamic equilibrium is reached when the rates of the forward and reverse reactions are equal. At equilibrium, the concentrations of reactants and products remain constant, but are unlikely to be equal.

To maximise the amount of products, chemists try to move the position of equilibrium in favour of the products. The next few pages show how this can be done.

Changing the concentration

Consider the equilibrium system that is present in a bottle of vinegar (dilute ethanoic acid):

$$CH_3COOH(aq) \rightleftharpoons CH_3COO^-(aq) + H^+(aq)$$
ethanoic acid molecules ethanoate ions hydrogen ions

As the system is at equilibrium, the rate of the forward reaction is equal to the rate of the reverse reaction and therefore the concentrations of the reactants and products are constant.

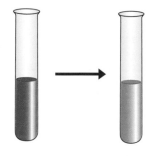

What happens if the concentration of ethanoate ions is changed? This can be done by adding solid sodium ethanoate to the equilibrium mixture.

The changes can be followed by adding universal indicator solution to the mixture.

The universal indicator is an orange–red colour in ethanoic acid. When solid sodium ethanoate is added, the universal indicator changes to a yellow–orange colour. This indicates that the solution has become less acidic, i.e. there are now fewer H^+ ions in the solution.

We can see from the equilibrium equation that if there are now fewer H^+ ions, then the position of equilibrium must have shifted to the left. There are now more CH_3COOH molecules and fewer H^+ ions.

Now consider the equilibrium shown in the equation below:

$$Cr_2O_7^{2-}(aq) + OH^-(aq) \rightleftharpoons 2CrO_4^{2-}(aq) + H^+(aq)$$
orange   colourless   yellow   colourless

The equilibrium mixture is a yellow–orange colour.

If sodium hydroxide is added (adding OH^- ions), the colour of the solution becomes more yellow, so the equilibrium position has shifted to the right.

If hydrochloric acid is added (adding H^+ ions), the colour of the solution becomes more orange, so the position of equilibrium has shifted to the left.

We can say that:

- adding a species which appears on the right-hand side of the equilibrium position shifts the position of equilibrium to the left

- adding a species which appears on the left-hand side of the equilibrium position shifts the position of equilibrium to the right.

It is also true that:

- removing a species which appears on the right-hand side of the equilibrium position shifts the position of equilibrium to the right

- removing a species which appears on the left-hand side of the equilibrium position shifts the position of equilibrium to the left.

DON'T FORGET

To shift the position of equilibrium to the right (to make more of the product) we can:
- add more of one of the substances on the left-hand side of the equation (the reactants), or
- remove the products as they are formed.

contd

Changing the pressure

Changes in pressure will only affect the position of equilibrium if a gas or gases are present.

Increasing the pressure shifts the position of equilibrium to the side of the equation which has the lower gas volume (lower number of moles of gas).

Decreasing the pressure shifts the position of equilibrium to the side of the equation which has the higher gas volume (higher number of moles of gas).

Consider the equilibrium shown in the equation below:

$$N_2O_4(g) \rightleftharpoons 2NO_2(g)$$

light yellow   dark brown

1 mol gas   2 mol gas

Increasing the pressure shifts the position of equilibrium backwards to the left because the volume of gas on the left-hand side is smaller than the volume of gas on the right-hand side. Therefore the colour of the gas mixture becomes lighter as the pressure is increased.

Decreasing the pressure shifts the position of equilibrium forward to the right because the volume of gas on the right-hand side is larger than the volume of gas on the left-hand side. Therefore the colour of the gas mixture becomes darker as the pressure is decreased.

Another example is the preparation of methanol from a mixture of carbon monoxide and hydrogen. The equilibrium equation for the reaction is:

$$CO(g) + 2H_2(g) \rightleftharpoons CH_3OH(g)$$

This equation shows three moles of gas on the left-hand side and only one mole of gas on the right-hand side.

Increasing the pressure shifts the position of equilibrium to the side with a lower volume of gas – that is, to the right. The preparation of methanol from carbon monoxide and hydrogen therefore gives a greater yield at higher pressures.

 ## THINGS TO DO AND THINK ABOUT

A reaction involving a gas or gases will only reach an equilibrium position if it is carried out in a closed container. If the container is open, then the gas(es) will escape and the system will not reach equilibrium.

Consider the equilibrium equation:

$$CaCO_3(s) \rightleftharpoons CaO(s) + CO_2(g)$$

- If this is carried out in an open container – for example, in a test-tube – then the carbon dioxide gas will escape into the air and the reverse reaction cannot take place. This means that the forward reaction will always be faster than the reverse reaction. Therefore equilibrium will not be reached and, eventually, all the calcium carbonate will have decomposed into calcium oxide. Equilibrium will only be reached if the reaction is carried out in a closed container from which no carbon dioxide is allowed to escape. The reverse reaction can then take place and equilibrium will be reached when the rate of the reverse reaction is the same as the rate of the forward reaction.

- If the reaction is carried out in a closed container and equilibrium has been reached, what effect will increasing the pressure have on the position of equilibrium?

  The equation $CaCO_3(s) \rightleftharpoons CaO(s) + CO_2(g)$ has zero moles of gas on the left-hand side and one mole of gas on the right-hand side. Increasing the pressure shifts the position of equilibrium to the side with the lower gas volume. As zero is less than one, then the position of equilibrium will shift to the left-hand side and there will be fewer molecules of carbon dioxide at this new equilibrium position.

MORE ABOUT SHIFTING THE EQUILIBRIUM POSITION

Changing the temperature

For a chemical system in equilibrium, an increase in temperature favours the endothermic reaction and a decrease in temperature favours the exothermic reaction.

Consider the equilibrium reaction in which ΔH has a positive value. This means that the forward reaction is endothermic and therefore the reverse reaction must be exothermic. This is shown above and below the equilibrium arrow:

$$N_2O_4(g) \underset{\text{exothermic}}{\overset{\text{endothermic}}{\rightleftharpoons}} 2NO_2(g) \qquad \Delta H \text{ is positive}$$
light yellow   dark brown

If a sealed test-tube containing this equilibrium mixture is placed in a beaker of hot water, the endothermic reaction is favoured and the position of equilibrium shifts to the right. The contents of the test-tube therefore become a darker brown.

If the same sealed test-tube containing this equilibrium mixture is placed in a beaker of iced water, the exothermic reaction is favoured and the position of equilibrium shifts to the left. The contents of the test-tube therefore become lighter in colour.

Effect of a catalyst on the position of equilibrium

Consider the potential energy diagram for the following reversible reaction at equilibrium:

$$A + B \rightleftharpoons C + D$$

If a catalyst is added, the rates of both the forward and the reverse reactions will be increased.

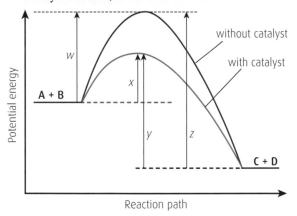

Reaction	Activation energy
Forward reaction, without catalyst	w
Forward reaction, with catalyst	x
Reverse reaction, without catalyst	z
Reverse reaction, with catalyst	y

The presence of the catalyst has lowered the activation energy for both the forward and reverse reactions by the same amount. Therefore, both the forward and reverse reactions have been speeded up equally and so the position of equilibrium has not been altered.

EQUILIBRIUM IN INDUSTRY: THE HABER PROCESS

The manufacture of ammonia by the Haber Process is a very important industrial reaction. The reactants are nitrogen and hydrogen gases. The reaction is reversible and, if the conditions are kept constant, the following equilibrium is reached:

$$N_2(g) + 3H_2(g) \rightleftharpoons 2NH_3(g) \qquad \Delta H = -92\,kJ\,mol^{-1}$$

Conditions are carefully chosen to provide a compromise between fast production, a high yield of ammonia and low costs. Factors which increase the rate of the forward reactions are therefore favourable.

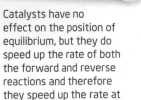

DON'T FORGET

In a system at equilibrium, an increase in temperature always favours the endothermic reaction, whereas a decrease in temperature always favours the exothermic reaction.

DON'T FORGET

Catalysts have no effect on the position of equilibrium, but they do speed up the rate of both the forward and reverse reactions and therefore they speed up the rate at which equilibrium is reached.

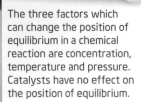

DON'T FORGET

The three factors which can change the position of equilibrium in a chemical reaction are concentration, temperature and pressure. Catalysts have no effect on the position of equilibrium.

contd

Pressure

As both the reactants and products are gases and the number of moles of the gaseous products is less than the number of moles of the gaseous reactants, a high pressure will shift the position of equilibrium to the side of the products. The pressure chosen is between 50 and 250 atm. Higher pressures could be used, but the relative increase in yield would not justify the increased cost.

Temperature

As the forward reaction is exothermic (ΔH is negative), lowering the temperature will favour the forward reaction. However, the reaction is very slow at low temperatures and a compromise between a slow reaction at low temperatures and a low yield at high temperatures has to be reached. As a result, the Haber Process is carried out at a moderate temperature of approximately 400–500°C. If the temperature is much higher than this, the faster reaction will not compensate for the much lower yield.

Recycling of unused gases

The yield in the Haber Process is, in fact, only about 14%. The mixture produced of ammonia gas and the unreacted nitrogen and hydrogen gases is then cooled down; the ammonia gas condenses into ammonia liquid, which is then tapped off. The unreacted nitrogen and hydrogen are then recirculated back or recycled through the reaction process.

Removal of product

Removal of the ammonia gas formed ensures that the rate of the reverse reaction is never equal to the rate of the forward reaction and therefore the system is not allowed to reach equilibrium. It is not in the manufacturer's interests for the ammonia formed to change back into nitrogen and hydrogen.

Catalyst

A catalyst lowers the activation energy of both the forward and reverse reactions. Using a catalyst therefore increases the rates of both the forward and reverse reactions. This allows the process to be carried out more quickly at a lower temperature. An iron catalyst is used in the Haber Process.

In industry, the best practice is usually a compromise between factors such as yield, rate of reaction and operating costs. A catalyst will speed up the reaction at a lower temperature and so reduce costs, but will have no effect on the yield. However the product will be made more quickly.

VIDEO LINK

Learn more about the Haber process by watching the videos at www.brightredbooks.net

 THINGS TO DO AND THINK ABOUT

Another important industrial process is the Contact Process for the manufacture of sulfuric acid. The equation for one stage in the reaction is:

$$2SO_2(g) + O_2(g) \rightleftharpoons 2SO_3(g)$$

In this equilibrium reaction, the reactants and product are again all gases. The total number of moles of gaseous reactants is greater than the number of moles of gaseous product. Therefore an increase in pressure would shift the position of equilibrium to the right.

However, the Contact Process is carried out at a pressure only slightly above atmospheric pressure, rather than the much higher pressure used in the Haber Process.

The reason that high pressure is used in the Haber Process is to increase the very low yield. The yield in the Contact Process is already over 90% at atmospheric pressure. Therefore the extra costs of a high pressure would not be justified as the yield would only increase marginally.

ONLINE TEST

How well have you learned about equilibria? Take the test at www.brightredbooks.net

CHEMICAL ENERGY 1

DON'T FORGET

$\Delta H = H_p - H_R$ where H_p is the enthalpy, or energy content, of the products and H_R is the enthalpy, or energy content, of the reactants.

DON'T FORGET

In an exothermic reaction, the temperature of the surroundings increases because chemical energy is being changed into heat energy. Therefore ΔH is negative. In an endothermic reaction, the temperature of the surroundings decreases because heat energy is being changed into chemical energy. Therefore ΔH is positive.

DON'T FORGET

The enthalpy change is measured in kilojoules (kJ) and, when worked out for one mole of the substance, the units are kJ mol⁻¹.

DON'T FORGET

Enthalpy of combustion values are always negative and are always quoted for one mole of the substance being burned.

DON'T FORGET

$E_h = cm\Delta T$
Remember that c, m and ΔT all refer to the **water** and that the mass of water is in kg. This relationship is given on page 4 of the Data Booklet.

ENTHALPY CHANGES

Enthalpy was first mentioned and defined on page 8. We can think of enthalpy as a measure of the chemical energy stored in a substance. It is given the symbol H. The chemical energy varies from substance to substance, although it is not possible to measure the absolute enthalpy of a substance. What can be measured, however, is the change in enthalpy, ΔH, which takes place in a chemical reaction.

Importance of enthalpy changes

Most chemical reactions are exothermic. Typical exothermic reactions are combustion and neutralisation. It is very important that chemists can predict accurately the amount of heat energy taken in or given out in industrial processes.

If the reactions are endothermic, heat energy will need to be supplied to keep the reaction going once it has started. This will be expensive.

If the reactions are exothermic, the heat produced may have to be removed to prevent the temperature rising and becoming dangerous. The heat energy can often be used elsewhere in the process so it is not wasted.

Enthalpy of combustion

The **enthalpy of combustion** of a substance is the enthalpy change when **one mole of the substance burns completely in oxygen**.

Enthalpy of combustion values are given in the Data Booklet (page 10). The values are always negative because combustion is an exothermic reaction. The units are in kJ mol⁻¹ because it is the energy value for one mole of the substance burning completely in oxygen.

The enthalpy of combustion values of some substances are given in the table on the right.

Balanced chemical equations which match the enthalpy of combustion values must always contain one mole of the substance being burned. This may mean that these equations contain fractions of moles of other substances, particularly oxygen. Look at the examples below:

Substance	Enthalpy of combustion ΔH_c/kJmol⁻¹
Hydrogen	−286
Carbon	−394
Methane	−891
Ethane	−1561
Propane	−2219
Butane	−2878
Methanol	−726
Ethanol	−1367

1 mole of hydrogen burning	$H_2(g) + \frac{1}{2}O_2(g) \rightarrow H_2O(l)$
1 mole of ethane burning	$C_2H_6(g) + 3\frac{1}{2}O_2(g) \rightarrow 2CO_2(g) + 3H_2O(l)$
1 mole of methanol burning	$CH_3OH(l) + 1\frac{1}{2}O_2(g) \rightarrow CO_2(g) + 2H_2O(l)$

It is perfectly acceptable to have fractions in balanced chemical equations as these are fractions of moles, not fractions of molecules.

Calculating enthalpy changes from experimental results

One way of measuring the enthalpy change for a chemical reaction is to arrange for the energy to be transferred to or from a known volume or mass of water. If the reaction is exothermic, then energy will be transferred to the water and the temperature of the water will increase. If it is endothermic, energy will be transferred from the water and its temperature will decrease.

The energy transferred to or from the **water** is:

$E_h = cm\Delta T$ where c = specific heat capacity of **water** (4·18 kJ kg⁻¹ °C⁻¹ as given in the Data Booklet), m = mass of **water** in kg and ΔT = change in temperature of **water** in °C.

This gives a value in kJ and the enthalpy change is then calculated per mole of substance, giving a value in kJ mol⁻¹.

contd

Calculating the enthalpy of combustion from experimental results

Example:

Using the apparatus shown, it was found that when 0·34 g of ethanol was burned in air, the temperature of 100 cm³ of water rose by 10°C. Calculate the enthalpy of combustion of ethanol.

From the Data Booklet (page 22), c = 4·18 kJ kg⁻¹ °C⁻¹.

The volume of water = 100 cm³, therefore the mass of water, m = 100 g = 0·100 kg; ΔT = 10°C.

The heat taken in by the water, $E_h = cm\Delta T$ = 4·18 × 0·100 × 10 = 4·18 kJ.

This is the heat given out by 0·34 g of ethanol, C_2H_5OH.

One mole of ethanol = 46 g, so the number of moles of ethanol used, $n = \frac{mass}{GFM} = \frac{0·34}{46}$ = 0·0074 mol.

0·0074 mol gives out 4·18 kJ so 1 mol gives out $\frac{4·18}{0·0074}$ = 565 kJ mol⁻¹.

Combustion is exothermic and so ΔH = −565 kJ mol⁻¹ from these experimental results.

This is much lower than the value given in the Data Booklet. The reasons for this include:
- heat loss to the surroundings, including the air, the copper can and the thermometer
- the ethanol is burning in air rather than in pure oxygen and so undergoes incomplete combustion
- the mass of ethanol may be incorrect due to some ethanol evaporating.

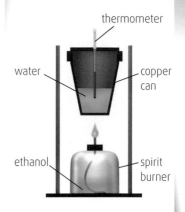

thermometer

water

copper can

ethanol

spirit burner

 THINGS TO DO AND THINK ABOUT

1 When 0·14 g of methanol combusts completely, 3·04 kJ of heat is given out. Calculate the enthalpy of combustion of methanol.

2 From the experimental results below, calculate the enthalpy of combustion of propan-1-ol and give two reasons why your calculated value is less than the one given in the Data Booklet:

Mass of empty copper can: 62·3 g

Mass of copper can + water: 164·3 g

Temperature of water in can at start: 18·0°C

Highest temperature that the water in the can reached: 28·5°C

Mass of spirit burner containing propan-1-ol before combustion: 24·36 g

Mass of spirit burner containing propan-1-ol after combustion: 24·14 g

3 Look at the diagram of the calorimeter below. It is sometimes known as a 'bomb' calorimeter. Suggest how using this to determine the enthalpies of combustion of alcohols will give accurate values.

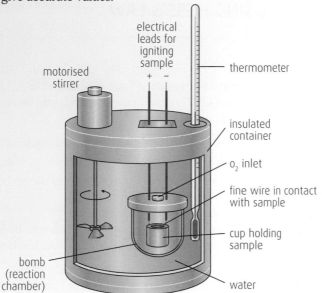

motorised stirrer

electrical leads for igniting sample
+   −

thermometer

insulated container

O₂ inlet

fine wire in contact with sample

cup holding sample

bomb (reaction chamber)

water

CHEMICAL ENERGY 2

CALCULATING ENTHALPY CHANGES USING HESS'S LAW

When considering ways to make new products, chemists are very interested in finding out the enthalpy change for the reaction. Endothermic reactions can be expensive if heat needs to be added, whereas exothermic reactions can be potentially dangerous if large amounts of heat are released very quickly. Hess's law gives chemists a way of predicting the enthalpy change for a new reaction using values for well-established reactions obtained from data books.

What is Hess's Law?

Hess's Law states that the enthalpy change for a chemical reaction is independent of the route taken.

This statement follows on from the law of Conservation of Energy – that is, energy cannot be created or destroyed.

Imagine a reaction which can take place by two different routes:
B can be made directly from A (route 1) or via C (route 2).
According to Hess's Law, $\Delta H_1 = \Delta H_2 + \Delta H_3$.

Manipulating chemical equations

Calculations are usually carried out by first writing out the balanced chemical equation for the enthalpy change you are trying to calculate.

Other chemical equations are then manipulated to give the desired equation. These equations can be reversed or multiplied. Whatever operation is carried out on the equation, then the same operation has to be carried out on the value of the enthalpy change.

For example, the equation for the combustion of hydrogen is:

$$H_2(g) + \tfrac{1}{2}O_2(g) \rightarrow H_2O(l) \qquad \Delta H = -286\,\text{kJ mol}^{-1}$$

If, in a calculation, this equation has to be doubled, then the value for the enthalpy change will also have to be doubled to −572 kJ.

If the equation has to be reversed to:

$H_2O(l) \rightarrow H_2(g) + \tfrac{1}{2}O_2(g)$ then the value of ΔH becomes +286 kJ.

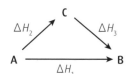

DON'T FORGET

Whatever operation is carried out on the equation, then that same operation must be carried out on the value for the enthalpy change. If the equation is reversed, then the sign of the enthalpy change value must also be changed.

CALCULATIONS USING HESS'S LAW

The equation which represents the formation of ethane from carbon and hydrogen is:

$$2C(s) + 3H_2(g) \rightarrow C_2H_6(g) \qquad \Delta H = ?$$

In practice, carbon and hydrogen do not react to form ethane and so ΔH for this reaction cannot be determined experimentally. However, a calculation based on Hess's Law can be used to determine the value of the enthalpy change.

The enthalpies of combustion of carbon, hydrogen and ethane have been determined experimentally and are given in the Data Booklet on page 10.

This can be used to calculate ΔH for $2C(s) + 3H_2(g) \rightarrow C_2H_6(g)$ as shown below.

Step 1

Write the balanced chemical equations for the combustion of carbon, hydrogen and ethane and note the enthalpy of combustion values from the Data Booklet.

(1) $C(s) + O_2(g) \rightarrow CO_2(g)$ $\qquad \Delta H = -394\,\text{kJ mol}^{-1}$
(2) $H_2(g) + \tfrac{1}{2}O_2(g) \rightarrow H_2O(l)$ $\qquad \Delta H = -286\,\text{kJ mol}^{-1}$
(3) $C_2H_6(g) + 3\tfrac{1}{2}O_2(g) \rightarrow 2CO_2(g) + 3H_2O(l)$ $\qquad \Delta H = -1561\,\text{kJ mol}^{-1}$

contd

Step 2

Write a balanced equation representing the desired enthalpy change:

$$2C(s) + 3H_2(g) \rightarrow C_2H_6(g) \qquad \Delta H = ?$$

This is the target equation and we can use Hess's Law to reach this equation from the three equations in Step 1.

Step 3

This involves using equations (1), (2) and (3) to reach the target equation.

In the target equation, there are two carbons on the left-hand side and in equation (1) there is one carbon on the left-hand side. Equation (1) therefore has to be multiplied by two and so does the enthalpy change value. Similarly, equation (2) has to be multiplied by three as the target equation has $3H_2$ on the left-hand side and equation (2) has only one H_2 on the left-hand side. The enthalpy change value will also have to be multiplied by three.

The target equation has $C_2H_6(g)$ on the right-hand side, whereas equation (3) has $C_2H_6(g)$ on the left-handside. Equation (3) has to be reversed and the sign of the enthalpy change value changed.

Putting these changes into place:

(1) × 2: $2C(s) + 2O_2(g) \rightarrow 2CO_2(g)$ $\Delta H = -394 \times 2 = -788\,kJ$
(2) × 3: $3H_2(g) + 1\frac{1}{2}O_2(g) \rightarrow 3H_2O(l)$ $\Delta H = -286 \times 3 = -858\,kJ$
Reverse (3): $2CO_2(g) + 3H_2O(l) \rightarrow C_2H_6(g) + 3\frac{1}{2}O_2(g)$ $\Delta H = +1561\,kJ$

Adding up these three equations gives:

$$2C(s) + 3H_2(g) + 3\tfrac{1}{2}O_2(g) + 2CO_2(g) + 3H_2O(l) \rightarrow 2CO_2(g) + 3H_2O(l) + C_2H_6(g) + 3\tfrac{1}{2}O_2(g)$$

Removing $3\frac{1}{2}O_2(g) + 2CO_2(g) + 3H_2O(l)$ which appear on both sides of the equation, gives:

$2C(s) + 3H_2(g) \rightarrow C_2H_6(g)$ which is the target equation.

As adding these three equations together gives the target equation, then ΔH for this reaction will be the sum of the enthalpy change values for the three equations. Therefore:

$$\Delta H = -788 - 858 + 1561 = -85 \text{ kJ mol}^{-1}$$

This shows how Hess's Law can be used to calculate the enthalpy change for a reaction which cannot be carried out in practice.

DON'T FORGET

Enthalpy of combustion values are given on page 10 of the Data Booklet.

THINGS TO DO AND THINK ABOUT

If there is an oxygen molecule in the target equation, it is obviously not possible to write an equation for the combustion of oxygen. However, oxygen will be present in the equations for the burning of other substances and when the equations are written and then manipulated correctly, the number of moles of oxygen molecules in the target equation will be correct.

For example, if you have to calculate the enthalpy of formation of ethanol from the enthalpy of combustion data in the Data Booklet, you would use the equations:

(1) $C(s) + O_2(g) \rightarrow CO_2(g)$ $\Delta H = -394\,kJ\,mol^{-1}$
(2) $H_2(g) + \frac{1}{2}O_2(g) \rightarrow H_2O(l)$ $\Delta H = -286\,kJ\,mol^{-1}$
(3) $C_2H_5OH(l) + 3O_2(g) \rightarrow 2CO_2(g) + 3H_2O(l)$ $\Delta H = -1367\,kJ\,mol^{-1}$

The target equation for the enthalpy of formation of ethanol from its elements is:

$$2C(s) + 3H_2(g) + \tfrac{1}{2}O_2(g) \rightarrow C_2H_5OH(g) \qquad \Delta H = ?$$

You will find that if you multiply equation (1) by two and equation (2) by three and then reverse equation (3) and add all these up, you will obtain the target equation, including the $\frac{1}{2}O_2(g)$ on the left-hand side as required. Carrying out the same operations on the enthalpy change values will give the enthalpy of formation of ethanol. Try this for yourself. You should get the answer $\Delta H = -279$ kJ mol^{-1}.

ONLINE TEST

Test yourself on chemical energy online at www.brightredbooks.net

CHEMICAL ENERGY 3

BOND ENTHALPY

What is the bond enthalpy?

The bond enthalpy (or **bond dissociation enthalpy**) is the energy required to break one mole of chemical bonds in the gaseous state to form two moles of gaseous atoms. In general, for a diatomic molecule XY, it is the energy required for the process:

$$X-Y(g) \rightarrow X(g) + Y(g)$$

Energy is required to break a bond and so, to comply with Hess's Law, the same amount of energy must be given out when that bond is made. This means that bond-breaking is endothermic (ΔH = positive) and bond-making is exothermic (ΔH = negative).

Tables in the Data Booklet (page 10) give bond enthalpies for the most common bonds you will encounter.

One table has the bond enthalpies for the bonds inside diatomic molecules. For example, if we consider the diatomic molecule hydrogen fluoride, you can see that the bond enthalpy for the H−F bond is given in the Data Booklet as 570 kJ mol^{-1}. This means that 570 kJ of energy is required to break one mole of bonds and that 570 kJ of energy will be given out when one mole of bonds is formed. The only substance in which the H−F bond exists is hydrogen fluoride. This is true for all the bonds in this table – that is, the bonds in this table will each only exist in one substance and so these are very accurately determined values.

The other table shows the **mean** or average bond enthalpies for bonds that can be found in more than one substance. For example, the C−H bond is found in alkanes, alkenes, alcohols, aldehydes and in all the other compounds you will encounter in Nature's Chemistry. The C−H bond may have a different bond enthalpy in methane than in the many other compounds containing C−H bonds. Therefore the mean value of the bond enthalpy of the C−H bond in all its compounds is in the Data Booklet and this is given as 412 kJ mol^{-1}.

The mean bond enthalpy for the O−H bond is given as 463 kJ mol^{-1} in the Data Booklet.

Therefore ΔH for the reaction $H_2O(g) \rightarrow 2H(g) + O(g)$ is $2 \times 463 = 926$ kJ as two moles of O−H bonds are being broken for every one mole of $H_2O(g)$.

When doing calculations using bond enthalpy data, you must consider all the bonds to be broken and all the bonds to be made. It is always best to draw out the full structural formulae of the reactants and products so that you can see all the bonds which are broken and made.

Worked example 1

Sometimes it is not necessary to use the bond enthalpies for all the bonds in the structures. For example, if we consider the reaction:

$$Cl_2(g) + C_2H_6(g) \rightarrow C_2H_5Cl(g) + HCl(g)$$

Writing this out again using structural formulae to show all the bonds:

You can see that, in this reaction, not all the bonds have to be broken. Therefore you may consider that only the Cl−Cl bond and one C−H bond have to be broken. The bonds that are made are one C−Cl bond and one H−Cl bond.

contd

DON'T FORGET

Bond-making is exothermic and bond-breaking is endothermic.

DON'T FORGET

$H-F(g) \rightarrow H(g) + F(g)$,
ΔH = 570 kJ mol^{-1};
$H(g) + F(g) \rightarrow H-F(g)$,
ΔH = -570 kJ mol^{-1}.

DON'T FORGET

Draw out the full structural formulae of all the reactants and products so you can see which bonds have to be broken and which ones need to be made.

Using values from the Data Booklet, and remembering that for bond-breaking, ΔH is positive and for bond-making, ΔH is negative, then the overall enthalpy change is $243 + 412 - 338 - 432 = \textbf{-115 kJ mol}^{-1}$.

Worked example 2

Calculate the enthalpy change in the following reaction using bond enthalpy data from the Data Booklet.

$CH_4(g) + 2O_2(g) \rightarrow CO_2(g) + 2H_2O(g)$

Using structural formulae this becomes:

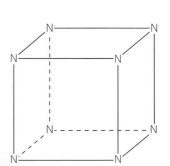

Bond-breaking (ΔH positive)

$4 \times (C-H)$, $\Delta H = 4 \times 412 = 1648$ kJ

$2 \times (O=O)$, $\Delta H = 2 \times 498 = 996$ kJ

Bond-making (ΔH negative)

$2 \times (C=O)$, $\Delta H = 2 \times -743 = -1486$ kJ

$4 \times (O-H)$, $\Delta H = 4 \times -463 = -1852$ kJ

Therefore the overall enthalpy change is $= 1648 + 996 - 1486 - 1852 = -694$ kJ mol^{-1}.

As some of the bond enthalpy values given in the Data Booklet are mean values, the enthalpy changes calculated from bond enthalpy values often differ from those determined experimentally.

THINGS TO DO AND THINK ABOUT

1 Using the enthalpy of combustion data from the Data Booklet, calculate the enthalpy change for the following reaction:

$C_2H_4(g) + H_2(g) \rightarrow C_2H_6(g)$

2 Using bond enthalpy data from the Data Booklet, calculate the enthalpy change for the same reaction as in Question 1.

3 Suggest why the values calculated in Questions 1 and 2 are different.

4 Use bond enthalpy data to calculate the enthalpy change for:

$H_2(g) + \frac{1}{2}O_2(g) \rightarrow H_2O(g)$

The value given in the Data Booklet for the enthalpy of combustion of hydrogen is -286 kJ mol^{-1}. Suggest two reasons why the answer you have calculated is not the same value.

5 Nitrogen normally exists as diatomic molecules. It has been suggested that it might be possible to prepare a new molecular form of nitrogen, N_8, with the following cubic structure:

The N−N bond enthalpy is 163 kJ mol^{-1} and the N≡N bond enthalpy is 945 kJ mol^{-1}.

a Calculate ΔH for the reaction $4N_2(g) \rightarrow N_8(g)$.

b Suggest why $N_8(g)$ has never been made successfully.

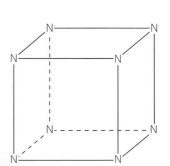

OXIDISING AND REDUCING AGENTS 1

WHAT ARE OXIDISING AND REDUCING AGENTS?

Oxidation is the loss of electrons (**OIL** – **O**xidation **I**s **L**oss). An **oxidising agent** is a substance which causes the oxidation of another reactant to take place. As oxidation is the loss of electrons, then an oxidising agent must be a substance which accepts electrons. The oxidising agent is itself reduced during a redox reaction.

Reduction is the gain of electrons (**RIG** – **R**eduction **I**s **G**ain). A **reducing agent** is a substance which causes the reduction of another reactant to take place. As reduction is the gain of electrons, then a reducing agent must be a substance which donates electrons. The reducing agent is itself oxidised during a redox reaction.

A typical **redox** (**red**uction/**ox**idation) reaction is magnesium burning in oxygen. Let us look at the balanced equation to see if we can identify the oxidising and reducing agents:

$$2Mg(s) + O_2(g) \rightarrow 2MgO(s)$$

To do this, we need to split up the redox equation into its oxidation and reduction ion–electron equations. This is straightforward provided you recognise that magnesium oxide is ionic and is made up of Mg^{2+} and O^{2-} ions. The Mg atoms have been converted into Mg^{2+} ions and so the relevant ion–electron equation must be:

$$Mg(s) \rightarrow Mg^{2+}(s) + 2e^-$$

We can see that the Mg atoms have been oxidised and therefore the O_2 molecules must have been reduced:

$$O_2(g) + 4e^- \rightarrow 2O^{2-}(s)$$

As the oxygen molecules are accepting electrons, they must be the oxidising agents. This implies that the magnesium atoms must be the reducing agents and this is confirmed by the fact that they are donating electrons.

Elements as oxidising and reducing agents

The **electrochemical series** is a list of metals and non-metals arranged in order of how easily they will lose electrons. An electrochemical series can be seen on page 12 in the Data Booklet.

Each reaction is shown as a reduction (a gain of electrons) and so the electrons are shown on the left-hand side of the arrow. The arrow is an equilibrium double arrow. This tells us that these reactions can also be reversed to become oxidation reactions.

Metal elements such as lithium and potassium have low electronegativity values and so tend to lose electrons readily to form positive ions. This is oxidation and therefore these so-called electropositive metals are good reducing agents. You will see that they are found at the top right-hand side of the electrochemical series and this is where the best reducing agents are located. The strongest reducing agents are from Group 1 in the Periodic Table. So, looking at the electrochemical series, the most effective reducing agent is Li(s).

The elements shown in red, the Group I metals, are the best reducing agents.

Non-metallic elements such as fluorine and chlorine have high electronegativity values and so they tend to gain electrons, forming negative ions. This is reduction and therefore non-metals tend to be good oxidising agents. The best oxidising agents are found at the bottom left of the electrochemical series. The strongest oxidising agents come from Group 7 in the Periodic Table and the most effective oxidising agent is $F_2(g)$.

contd

DON'T FORGET

An oxidising agent is an electron acceptor and a reducing agent is an electron donor.

DON'T FORGET

Reducing agents are on the right-hand side of the electrochemical series and the strongest reducing agents are at the top right-hand side of the electrochemical series.

The elements shown in green, which are all non-metals, act as oxidising agents.

Compounds as oxidising and reducing agents

Compounds can also act as oxidising and reducing agents. For example, hydrogen peroxide, H_2O_2, is a strong oxidising agent and carbon monoxide, CO, is a very good reducing agent and can reduce some metal oxides to the metal.

For some ionic compounds, it is the ions within them that are the oxidising or reducing agents. Sometimes these ions are group ions. Looking at the electrochemical series you can see that the group ions, such as permanganate (MnO_4^-) and dichromate ($Cr_2O_7^{2-}$), lie near the bottom right of the table. This means that they are very good oxidising agents. You will also see, if you look carefully at their ion–electron equations, that there are also hydrogen ions, H^+, on the left-hand side of their equations. This tells us that they only function as effective oxidising agents in acidic solutions where there are abundant hydrogen ions.

Travelling further up the electrochemical series table, but on the right-hand side, you will find the sulfite ion, SO_3^{2-}(aq). As it is on the right-hand side, it is a reducing agent. The sulfite ion is a good example of a useful reducing agent.

Uses of oxidising agents

Oxidising agents are effective in killing fungi and bacteria and can inactivate some viruses. A major use of the oxidising agent potassium permanganate is as a steriliser – its strong oxidising properties make it an effective disinfectant. It can be used to treat fungal infections, such as athlete's foot. It can also be used to extend the shelf life of fruit. This is because ripening fruit gives off ethene gas, which, in turn, causes other fruit to ripen. Potassium permanganate reacts with ethene so, when fruits are shipped from abroad, their containers contain porous material impregnated with potassium permanganate to remove any ethene from the air. This is commercially important in extending the shelf life of fruit, flowers and vegetables.

Bleaching clothes and hair is usually an oxidation process. Many coloured compounds change their molecular structure when oxidised and become colourless. Obviously it is important that the oxidising agents used to bleach clothes and hair are not too concentrated. A dilute solution of hydrogen peroxide is sometimes used to bleach hair, giving rise to the term 'peroxide blondes'.

THINGS TO DO AND THINK ABOUT

1   Consulting the Data Booklet, write the appropriate ion–electron equations for:

   a   chlorine gas acting as an oxidising agent

   b   acidified permanganate ions in solution acting as an oxidising agent

   c   solid zinc acting as a reducing agent

   d   sulfite ions in solution acting as a reducing agent

   e   acidified dichromate ions in solution acting as an oxidising agent

2   If you look for Fe^{2+}(aq) in the electrochemical series in the Data Booklet, you will find it is present on the right-hand side near the foot of the table and also higher up on the left-hand side of the table. How can this be? The reason is that Fe^{2+}(aq) ions can act both as a reducing agent and as an oxidising agent. When acting as a reducing agent, the Fe^{2+}(aq) ions become oxidised to Fe^{3+}(aq) ions. When acting as an oxidising agent, the Fe^{2+}(aq) ions are reduced to solid Fe metal.

DON'T FORGET

Oxidising agents are on the left-hand side of the electrochemical series and the strongest oxidising agents are at the bottom left-hand side of the electrochemical series.

DON'T FORGET

Acidified permanganate ions and acidified dichromate ions and hydrogen peroxide are useful oxidising agents.

DON'T FORGET

The sulfite ion in solution and carbon monoxide gas are useful reducing agents.

DON'T FORGET

Many substances that are reducing agents or oxidising agents are shown in the table on page 12, in the Data Booklet. However, carbon monoxide and hydrogen peroxide are not in the table and you need to remember that CO(g) is a good reducing agent and that H_2O_2(l) is a good oxidising agent.

 ONLINE

Look online at www.brightredbooks.com to see further definitions of oxidising and reducing agents.

 ONLINE TEST

Head to www.brightredbooks.net and take the 'Oxidising and reducing agents' test.

OXIDISING AND REDUCING AGENTS 2

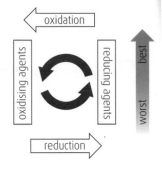

REDOX EQUATIONS AND PREDICTIONS

Ion-electron equations

Ion–electron equations are used to describe the oxidation and reduction processes that make up a redox reaction. In an oxidation ion–electron equation, the electrons appear on the products side because oxidation involves a loss of electrons. In a reduction ion–electron equation they are on the reactants side as reduction involves a gain of electrons. In any ion–electron equation, the net charge on the reactants side must always equal that on the products side.

To illustrate these ideas, consider the reaction between iron(II) sulfate solution and an acidified potassium dichromate solution in which the iron(II) ions change into iron(III) ions and the dichromate ions change into chromium(III) ions. The potassium ions and sulfate ions are 'spectator' ions and do not take part in the reaction.

The iron(II) ions are converted into iron(III) ions: $Fe^{2+}(aq) \rightarrow Fe^{3+}(aq)$

To balance the charges, an electron has to be added to the product side, giving a net charge of 2+ on each side: $Fe^{2+}(aq) \rightarrow Fe^{3+}(aq) + e^-$

As the electrons appear on the products side, the iron(II) ions have been oxidised. This implies that the dichromate ions in the acidified potassium dichromate solution must have been reduced to chromium(III) ions: $Cr_2O_7^{2-}(aq) \rightarrow Cr^{3+}(aq)$

Working out the ion–electron equation for this particular reduction process is not quite as straightforward as that for the oxidation, but it can be achieved by applying the following rules:

1 Balance the non-oxygen element, in this case the chromium:

$Cr_2O_7^{2-}(aq) \rightarrow 2Cr^{3+}(aq)$

2 Balance the oxygen. Oxygen is needed on the product side and is introduced as water molecules. As the dichromate ions contain seven oxygen atoms, seven water molecules will be required:

$Cr_2O_7^{2-}(aq) \rightarrow 2Cr^{3+}(aq) + 7H_2O(l)$

3 Balance the hydrogen. Hydrogen is needed on the reactants side and is introduced as hydrogen ions. Fourteen H^+ ions are required:

$Cr_2O_7^{2-}(aq) + 14H^+(aq) \rightarrow 2Cr^{3+}(aq) + 7H_2O(l)$

4 Finally, balance the charges. As it stands, the net charge on the reactants side is 12+ (14+ from the 14 H^+ ions and 2– from the one $Cr_2O_7^{2-}$ ion). On the products side, the net charge is 6+ from the two Cr^{3+} ions. To balance the charge, we introduce electrons. Six are needed on the reactants side to make the net charge on each side 6+:

$Cr_2O_7^{2-}(aq) + 14H^+(aq) + 6e^- \rightarrow 2Cr^{3+}(aq) + 7H_2O(l)$

The electrons appear on the reactants side, confirming that it is a reduction process.

As hydrogen ions are needed for dichromate ions to be reduced, this explains why the potassium dichromate solution had to be acidified.

Thankfully, most ion–electron equations are found on page 12 of the Data Booklet. As stated earlier, they are all written as reductions and so to write an oxidation ion–electron equation all you have to do is reverse the relevant reduction equation. There may be occasions when you have to write an electron equation which does not appear in the Data Booklet. Then you must derive it from first principles as outlined above. Ion–electron equations can be combined to give **redox equations**.

contd

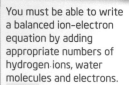

DON'T FORGET

You must be able to write a balanced ion-electron equation by adding appropriate numbers of hydrogen ions, water molecules and electrons.

Redox equations

Balanced equations for redox reactions can be written by combining the ion–electron equations for the oxidation and reduction processes. Consider the redox reaction between sodium sulfite solution and acidified potassium permanganate solution. The rule is to find the relevant ion–electron equations from the Data Booklet and then reverse the equation that is higher in the table so that it becomes oxidation. The two ion–electron equations are now:

oxidation: $$SO_3^{2-}(aq) + H_2O(l) \rightarrow SO_4^{2-}(aq) + 2H^+(aq) + 2e^-$$
reduction: $$MnO_4^-(aq) + 8H^+(aq) + 5e^- \rightarrow Mn^{2+}(aq) + 4H_2O(l)$$

The redox equation has to be balanced with respect to the electrons – that is, the number of electrons lost in the oxidation process must equal the number gained in the reduction process. This implies that we must multiply the oxidation equation by five and the reduction equation by two:

$$5SO_3^{2-}(aq) + 5H_2O(l) \rightarrow 5SO_4^{2-}(aq) + 10H^+(aq) + 10e^-$$
$$2MnO_4^-(aq) + 16H^+(aq) + 10e^- \rightarrow 2Mn^{2+}(aq) + 8H_2O(l)$$

With the electrons balanced, we add these ion–electron equations together giving:

$$5SO_3^{2-}(aq) + 5H_2O(l) + 2MnO_4^-(aq) + 16H^+(aq) \rightarrow 5SO_4^{2-}(aq) + 10H^+(aq) + 2Mn^{2+}(aq) + 8H_2O(l)$$

H^+ ions and H_2O molecules appear on both sides of the equation and these have to be cancelled down. We finally arrive at the balanced redox equation:

$$5SO_3^{2-}(aq) + 2MnO_4^-(aq) + 6H^+(aq) \rightarrow 5SO_4^{2-}(aq) + 2Mn^{2+}(aq) + 3H_2O(l)$$

Using the electrochemical series to predict whether reactions will take place

We know that the best oxidising agents are at the bottom left-hand side of the electrochemical series and that the best reducing agents are at the top of the right-hand column in the electrochemical series. This enables us to use the electrochemical series to predict if a redox reaction will take place.

What you must remember is that any species (ions, molecules or atoms) on the left-hand side will react with any species above it on the right-hand side. The lower equation stays as a reduction and the upper equation is reversed to become oxidation.

Consider $Fe^{2+}(aq)$, which, as you know, is to be found on both the right- and left-hand sides of the table. Oxidising agents are on the left-hand side of the arrows in the table, so when it is acting as an oxidising agent it will oxidise $Cr(s)$, $Zn(s)$ and all the other species above it on the right-hand side of the table. It will itself be reduced to $Fe(s)$.

Reducing agents are on the right-hand side of the arrows in the table, so when acting as a reducing agent $Fe^{2+}(aq)$ will reduce $Ag^+(aq)$, $Hg^{2+}(aq)$ and all the other species below it on the left-hand side of the table. It will itself be oxidised to $Fe^{3+}(aq)$.

 THINGS TO DO AND THINK ABOUT

We can now predict whether redox reactions will occur using this information. You should be able to predict that $Cl_2(aq)$ will oxidise $Br^-(aq)$ ions into $Br_2(l)$ molecules, but that $Br_2(l)$ molecules will not be able to oxidise $Cl^-(aq)$ ions into $Cl_2(aq)$.

$Na^+(aq)$ and $K^+(aq)$ are almost always spectator ions. They are on the left-hand side of the table and there are very few species above them on the right-hand side of the table.

Predict whether each of the following will result in a redox reaction. Where a redox reaction will take place, write the ion–electron equation for both the oxidation and reduction steps and then the overall redox equation.

a bromine solution added to potassium iodide solution
b zinc metal added to magnesium chloride solution
c zinc metal added to copper(II) chloride solution
d acidified potassium permanganate added to silver(I) nitrate solution
e acidified potassium dichromate added to sodium sulfite solution.

 **DON'T FORGET**

It is important to remember that neither electrons nor spectator ions/spectator molecules are present in the overall redox equation.

 **DON'T FORGET**

Any species on the left-hand side of the electrochemical series will oxidise any species above it on the right-hand side. Any species on the right-hand side of the electrochemical series will reduce any species below it on the left-hand side.

 ONLINE

The anti-clockwise rule, shown at the top of page 72, may help you here. For more information, head to www.brightredbooks.net

 VIDEO LINK

You can follow the link at www.brightredbooks.net to see some experiments involving halogen solutions and halide ions.

 ONLINE TEST

Head to www.brightredbooks.net and take the 'Oxidising and reducing agents' test.

CHEMICAL ANALYSIS 1

CHROMATOGRAPHY

What is chromatography?

Chromatography is a method used to analyse mixtures. It can be used to identify the substances present in a mixture and, in some cases, it can tell us how much of each substance is in the mixture.

There are different types of chromatography, but they all involve separating the components in a mixture by passing the mixture through a suitable medium in which the different components move at different speeds.

The medium is an adsorbent material and is often referred to as the **stationary phase**. The different components can be held to the stationary phase by van der Waals attractions. The stronger the van der Waals attractions formed between the component and the medium, the more the component is held up and therefore it travels more slowly through the medium. The weaker the bonds formed between the component and the medium, the less the component is held up and so it passes through the medium more quickly. As the different components move at different speeds through the medium, they are separated from each other.

Usually a liquid or gas is used to carry the mixture through the adsorbent stationary phase – because this moves through the medium, it is known as the **mobile phase**.

Chromatography works because it exploits the fact that different molecules experience different types and strengths of **intermolecular forces** as the mobile phase carries them through the stationary phase. The strengths and types of these intermolecular forces depend on the differences in the polarity and size of the molecules being separated.

Molecules that form stronger intermolecular forces with the mobile phase than with the stationary phase will move more quickly than other molecules that form stronger intermolecular forces with the stationary phase.

Using our knowledge of the different types of intermolecular forces, we can make predictions about which substances travel through the medium at a faster rate. Remember that 'like dissolves like'. If the mobile phase is polar and the stationary phase is non-polar, then we would expect any polar molecules to move further and faster than non-polar molecules as they are carried through the stationary phase by the mobile phase.

There are several types of chromatography. You are not expected to remember any details about them, but some information is given in the table below:

Type of chromatograpy	Mobile phase	Stationary phase
Paper chromatography	Liquid solvent	Paper
Thin-layer chromatography	Liquid solvent	Silica coating on a plastic film
Gas-liquid chromatography	Unreactive gas	Non-volatile liquid sticking to an unreactive solid

The finished result in both paper and thin-layer chromatography is known as a chromatogram and the different substances show up as spots. If the spots are colourless, then a compound known as a locating reagent can be sprayed on to the chromatogram to make the different compounds show up as coloured spots.

DON'T FORGET

There are different types of chromatography. Each type involves interactions between the components in the mixture to be separated and the mobile phase and the stationary phase.

DON'T FORGET

Molecular size and the polarity of molecules affect the speed at which molecules travel in chromatography; the differences in these properties are why components in a mixture separate out.

DON'T FORGET

In all types of chromatography, the components in a mixture are carried by the mobile phase through the stationary phase. The mobile phase and/or the stationary phase are different in the different types of chromatography.

contd

Gas–liquid chromatography is much more complicated and is carried out using special apparatus, as shown in the diagram on the right. The injection port is where the gas mixture or liquid mixture enters. The stationary phase is coiled inside an oven to vaporise any liquids that are to be separated and identified.

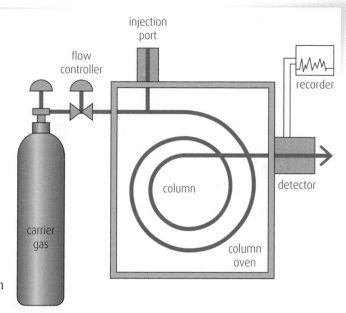

After separation, a detector produces a signal whenever a compound leaves the column. The results are then recorded on a graph.

The *y*-axis of the graph gives an indication of the amount of the component present and the *x*-axis gives the **retention time** of that component. The retention time is the time taken for a particular component to travel through the apparatus.

You will see examples of chromatographs below and on pages 76 and 77.

VIDEO LINK

Watch a video introducing paper chromatography at www.brightredbooks.net

VIDEO LINK

Watch the clip describing chromatography at www.brightredbooks.net

ONLINE TEST

Take the 'Chemical analysis' test at www.brightredbooks.net

THINGS TO DO AND THINK ABOUT

Chromatography is a useful analytical and forensic technique.

1. Chromatography can be used to find out if a substance is pure. For example, carrying out a thin-layer chromatography experiment using a pure substance results in only one spot on the developed chromatogram. If the substance had an impurity present, the impurity would show up as another spot. Consider the following example.

   An organic chemist is attempting to synthesise a fragrance compound by the following chemical reaction:

   compound X + compound Y → fragrance compound

   After one hour, a sample is removed and compared with pure samples of compounds X and Y using thin-layer chromatography.

   Which of the chromatograms on the right shows that the reaction has produced a pure sample of the fragrance compound?

2. Under a definite set of experimental conditions for chromatographic analysis, a given substance will always travel a fixed distance relative to the distance travelled by the solvent front. This ratio of distances is called the R_f value:

   $$R_f = \frac{\text{distance travelled by substance}}{\text{distance travelled by solvent front}}$$

   The diagram on the right shows how to calculate the R_f value of a substance from a chromatogram.

   As the R_f value is characteristic for any given compound (provided that the same stationary and mobile phases are used), it can provide evidence about the identity of the compound.

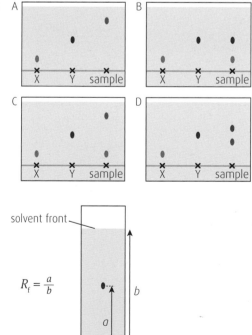

CHEMICAL ANALYSIS 2

MORE ON CHROMATOGRAPHY

Answering questions about chromatography in exams

In the Higher Chemistry exam you will not be asked to describe any type of chromatography in detail, but you may be asked to interpret and analyse the results of chromatography experiments.

Examples from recent examinations are given below:

1   Caffeine is added to some soft drinks. The concentration of caffeine can be found using chromatography.

A chromatogram for a standard solution containing 50 mg l^{-1} of caffeine is shown below.

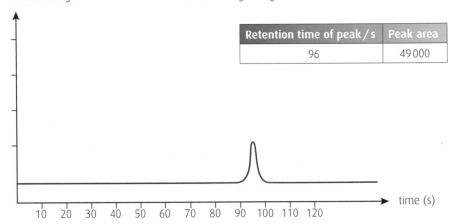

Retention time of peak / s	Peak area
96	49 000

Results from four caffeine standard solutions were used to produce the calibration graph below.

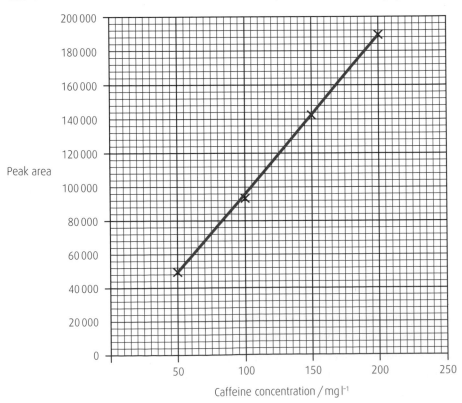

contd

Chromatograms for two soft drinks are shown below.

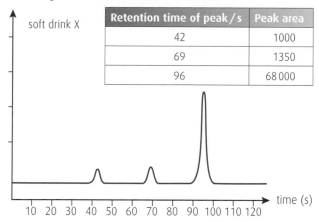

soft drink X

Retention time of peak / s	Peak area
42	1000
69	1350
96	68 000

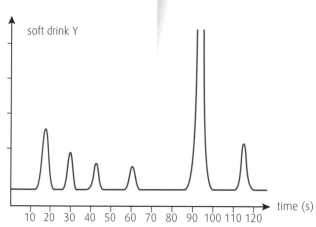

soft drink Y

a What is the caffeine content, in mg l⁻¹, of soft drink **X**?

Answer:

From the chromatogram on the previous page, the retention time for caffeine under these conditions is 96 s.

Looking at the chromatogram and table for soft drink **X**, the peak area is 68 000. When we match this against the graph, we can see that the caffeine concentration is 70 mg l⁻¹ in soft drink **X**.

b The caffeine content of the soft drink **Y** cannot be determined from its chromatogram.

What should be done to the sample of soft drink **Y** so that the caffeine content could be reliably calculated?

Answer:

You can see that the peak for caffeine in soft drink **Y** is too high and goes off the scale. The sample needs to be diluted or a smaller sample used.

VIDEO LINK

Watch more about the retention factor at www.brightredbooks.net

2 Diesel contains a mixture of non-polar molecules of different sizes.

Below are chromatograms recorded using a normal sample of diesel and a sample of diesel that has been heated until around 90% of the diesel has been evaporated.

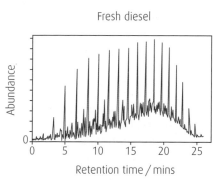

Fresh diesel

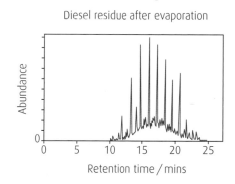

Diesel residue after evaporation

Explain how these chromatograms show that large molecules have longer retention times than small molecules in this type of chromatography.

Answer:

Smaller molecules will evaporate first and the chromatogram of the diesel residue after evaporation has no peaks with short retention times compared with the chromatogram of the fresh diesel.

THINGS TO DO AND THINK ABOUT

To find out about uses of chromatography in forensics head to www.brightredbooks.net and look at the article 'TLC the Forensic Way' and the video which shows how thin layer chromatography can be used in forensic chemistry.

ONLINE TEST

Take the 'Chemical analysis' test at www.brightredbooks.net

CHEMICAL ANALYSIS 3

VOLUMETRIC ANALYSIS

Standard solutions

A **standard solution** is a solution with an accurately known concentration. This will be different from many of the solutions you may have used in the chemistry laboratory. For example, bench hydrochloric acid usually has concentration of around 2 mol l^{-1}, but, in practice, could be anywhere between 1·5 and 2·5 mol l^{-1}. The concentrations of standard solutions are usually given to two or more significant figures to show that their concentration is accurately known. For example, the concentration of a standard solution would be written as 1·00 mol l^{-1} rather than 1 mol l^{-1} solution or perhaps 0·102 mol l^{-1} rather than 0·1 mol l^{-1}.

DON'T FORGET

A standard solution is a solution of accurately known concentration.

What happens in volumetric analysis?

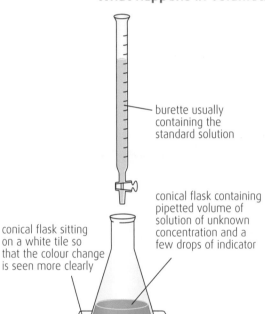

burette usually containing the standard solution

conical flask containing pipetted volume of solution of unknown concentration and a few drops of indicator

conical flask sitting on a white tile so that the colour change is seen more clearly

Volumetric analysis involves using a standard solution in a **quantitative reaction** to determine the concentration of another solution. A quantitative reaction is one in which the balanced equation is known. The volume of reactant solution required for the reaction to be completed is determined by **titration**.

In a titration, one of the solutions is added from a burette to a known volume of the other solution in a conical flask. The volume of this other solution is measured accurately using a **pipette**. Usually the solution of known concentration is in the burette.

If both the reactants and products are colourless, the point at which the reaction is just complete is usually detected by adding an **indicator** to the solution in the flask. A suitable indicator is one that changes colour when the reaction is just finished. The point at which the reaction is just complete is known as the end-point of the reaction. From the results, we can then calculate the accurate concentration of the other solution. The usual set up for a titration is shown in the diagram on the left.

DON'T FORGET

The solution from the burette is added until the indicator changes colour. The first titration is a rough experiment. More accurate titrations are carried until concordant results are obtained. The mean of the concordant results is the value used in the calculation.

The standard solution is added from the burette until there is a permanent colour change in the conical flask. This first titration is usually a rough experiment to give a rough indication of the volume of standard solution required. Titrations are repeated more carefully, with the solution from the burette being added drop by drop near the end-point, until concordant results are obtained. Concordant results are titration results that are within 0·2 cm^3 of each other. The mean value of the two concordant results is then used to calculate the concentration of the other solution. The rough titration value should not be used in the calculation. Another word for the titration value is the titre.

There are two types of titration with which you need to be familiar: acid–base titrations and redox titrations.

Acid–base titrations

DON'T FORGET

In any titration, the concentration of one of the solutions must be known accurately. This solution is the standard solution.

Acid–base titrations are based on neutralisation reactions in which the hydrogen ions from an acid react with the hydroxide ions from an alkali. Indicators used in acid–base titrations are often derived from plants. Indicators you may have used in the laboratory include methyl orange and phenolphthalein. A typical example of an acid–base titration is a measurement of the concentration of ethanoic acid in vinegar by titration with a standard solution of sodium hydroxide.

contd

Redox titrations

These are titrations based on redox reactions (see page 70). An oxidising agent is titrated against a reducing agent or vice versa. In redox reactions, electrons are transferred from the reducing agent to the oxidising agent.

Certain substances can be used as indicators in redox titrations, but an indicator is not required when potassium permanganate, a very good oxidising agent, is used as it acts as its own indicator.

Consider a redox titration involving purple potassium permanganate solution reacting with a standard solution of colourless sodium oxalate. The potassium and sodium ions are spectator ions and so the oxalate ions $(C_2O_4^{2-})$ are reacting with the permanganate ions (MnO_4^-). The overall redox equation is:

$$5C_2O_4^{2-} + 2MnO_4^- + 16H^+ \rightarrow 10CO_2 + 2Mn^{2+} + 8H_2O$$

If the colourless oxalate ions are in the burette, then the purple permanganate ions are in the conical flask. You can see from the redox equation that hydrogen ions are also required and so dilute sulfuric acid will have to be added to the conical flask.

As the titration gets underway, the colourless oxalate ions react with the permanganate ions, decreasing the concentration of the permanganate ions in the conical flask. The purple colour therefore becomes a lighter purple. As more oxalate ions are added, the purple colour becomes increasingly lighter until, at the end-point, the colour in the conical flask disappears completely.

It can be difficult to spot the exact moment the colour disappears and so sometimes the titration is carried out with the potassium permanganate in the burette and the sodium oxalate in the conical flask with the acid. This time the colour of the solution in the conical flask is colourless and the end-point is when the solution in the conical flask takes on a permanent pink/purple colour.

In titrations involving iodine molecules changing to iodide ions, or vice versa, starch solution can be used as the indicator.

DON'T FORGET

Potassium permanganate is a very useful reagent in redox titrations because it is a very good oxidising agent and it acts as its own indicator.

THINGS TO DO AND THINK ABOUT

Here is a calculation based on a redox titration.

A 50.0 cm³ sample of contaminated water containing chromate ions was titrated and found to require 27.4 cm³ of 0.0200 mol l⁻¹ iron(II) sulfate solution to reach the end-point.

The redox equation for the reaction is:

$$3Fe^{2+}(aq) + CrO_4^{2-}(aq) + 8H^+(aq) \rightarrow 3Fe^{3+}(aq) + Cr^{3+}(aq) + 4H_2O(l)$$

Calculate the chromate concentration, in mol l⁻¹, present in the sample of water.

Answer:

The sulfate ions are spectator ions and, as you can see, are not included in the overall redox equation. However, the redox equation does tell us that three moles of $Fe^{2+}(aq)$ ions react with one mole of $CrO_4^{2-}(aq)$ ions. Therefore the number of moles of $Fe^{2+}(aq)$ ions is:

$n = cV = 0.0200 \times 0.0274 = 0.000548$ mol

Therefore the number of moles of $CrO_4^{2-}(aq)$ ions is:

$n = \frac{0.000548}{3} = 0.00018267$ mol

This is the number of moles of $CrO_4^{2-}(aq)$ ions in 50.0 cm³ and so the concentration, c, of chromate ions, in mol l⁻¹, is $c = \frac{n}{V} = \frac{0.00018267}{0.050} = $ **0.00365 mol l⁻¹**.

ONLINE TEST

Take the 'Chemical analysis' test at www.brightredbooks.net

RESEARCHING CHEMISTRY

LABORATORY APPARATUS

CONICAL FLASKS

DON'T FORGET

Conical flasks are used in titrations because it is easy to swirl the contents without spilling any of the solution in the flask.

Conical flasks are ideal for carrying out reactions in which you need to mix the reactants together. Conical flasks have a narrow mouth, so it is easy to swirl the contents of the flask without the contents spilling out. They are also good for carrying out reactions in which a gas is given off because the narrow mouth helps to avoid any spray spilling over your bench.

A conical flask and its sectional diagram.

VOLUMETRIC FLASKS

Volumetric flasks are used for preparing standard solutions. A standard solution is a solution with an accurately known concentration.

Volumetric flasks come in different sizes so you can prepare different volumes of solution. A line is scratched or drawn on the thin neck of the flask to show the level to which the flask should be filled.

DON'T FORGET

Volumetric flasks are used to prepare standard solutions.

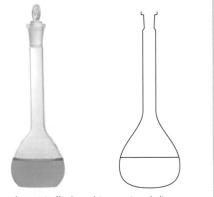

A volumetric flask and its sectional diagram.

MEASURING CYLINDERS

Measuring cylinders provide a very quick and easy way to measure out volumes of liquid reagents or solutions. They are accurate enough for use when measuring out the reactants required in a chemical reaction. However, they should not be used for analysis where an accurate result is required.

A measuring cylinder and its sectional diagram.

BEAKERS

DON'T FORGET

The volume markings on beakers only give a very rough measure of the volume in the beaker.

Beakers are used to hold reactant solutions just before use and for carrying out reactions.

Most beakers have lines on them to give an indication of the volume of the solution in the beaker, but these are not very accurate.

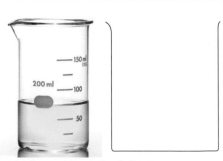

A beaker and its sectional diagram.

PIPETTES WITH SAFETY FILLERS

Pipettes measure out the volume of a solution very accurately. They come in a range of sizes. The solution is sucked up using a suction pump (called a 'safety filler') until the solution is level with a line drawn on the pipette. The pipette is then held over the reaction vessel and a button on the filler is pressed. This lets the solution drain out of the pipette. A small volume of liquid will always remain in the tip of the pipette. The pipette manufacturer has taken this into account, so do not attempt to blow this liquid out of the pipette.

Using a pipette with a safety filler and its sectional diagram.

ONLINE

Head to www.brightredbooks.net/ to try out using a virtual pipette filler!

BURETTES

Burettes are like an accurately graduated measuring cylinder with a tap at the bottom. They can be used either to measure how much of a solution is needed to react with another chemical, or to measure out 'odd' volumes (such as 23 cm³) for which no pipette is available. They are far more accurate than a measuring cylinder, but not as accurate as a pipette.

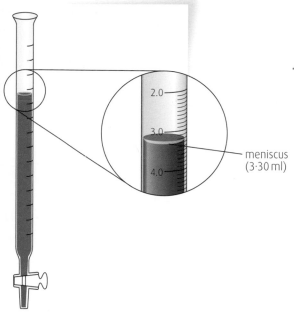

meniscus (3·30 ml)

A burette and a diagram showing how to accurately read the volume of liquid in a burette.

EVAPORATING BASINS

Evaporating basins are little porcelain or Pyrex bowls. They are used when you want to recover the solute from a solution. The solution is poured into the bowl and the solvent is allowed to evaporate. If required, the basin can be gently heated to boil the solvent off more quickly.

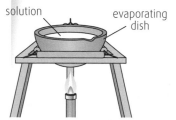

solution   evaporating dish

An evaporating basin and a diagram showing a basin being heated to boil off the solvent.

THINGS TO DO AND THINK ABOUT

When carrying out accurate work, it is important to rinse pipettes and burettes using the solution that is going to be measured in them. Why are volumetric flasks rinsed with distilled water before use but do not need to be dried?

TECHNIQUES 1

METHODS FOR COLLECTION OF GASES

Methods for collection of gases		Advantages	Disadvantages
Collection over water	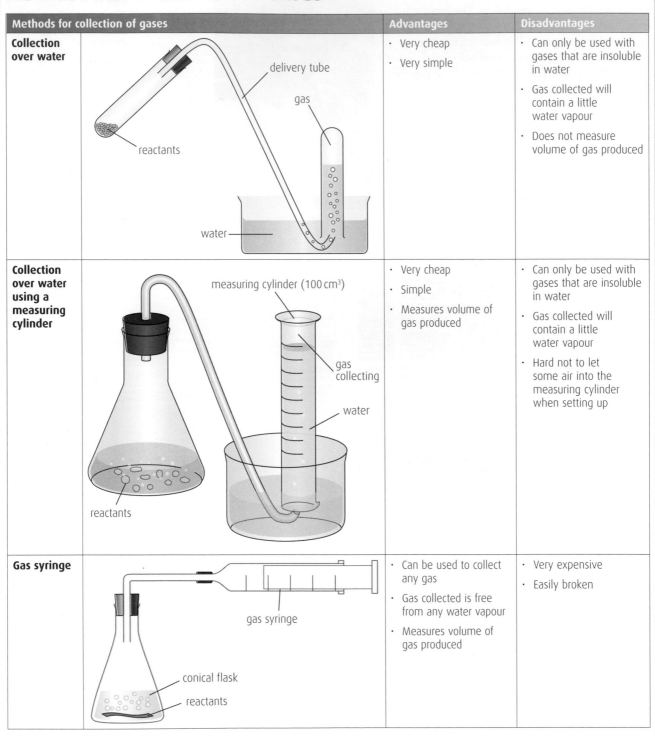	• Very cheap • Very simple	• Can only be used with gases that are insoluble in water • Gas collected will contain a little water vapour • Does not measure volume of gas produced
Collection over water using a measuring cylinder		• Very cheap • Simple • Measures volume of gas produced	• Can only be used with gases that are insoluble in water • Gas collected will contain a little water vapour • Hard not to let some air into the measuring cylinder when setting up
Gas syringe		• Can be used to collect any gas • Gas collected is free from any water vapour • Measures volume of gas produced	• Very expensive • Easily broken

SAFE HEATING METHODS

Safe heating methods		Advantages	Disadvantages
Bunsen burner	hottest part of the flame unburnt gas burner tube adjustable air hole gas supply	· Very cheap · Can achieve temperatures of over 1000°C · Instant change in temperature when you adjust the air hole	· **Must never be used to heat flammable compounds** · Does not apply heat evenly · Hard to control temperature accurately
Water-bath	test tube containing reaction mixture hot water beaker	· Can be used to heat flammable compounds · Can heat the entire surface of the reaction vessel · Can achieve very fine control of temperature	· Cannot be used to heat to above about 95°C · Takes time to reach any given temperature
Heating mantle or hot-plate	Heating mantle Hot-plate	· Can be used to heat flammable compounds · Can reach temperatures of about 400°C · Can achieve reasonable control of temperature	· Very expensive · Takes time to reach any given temperature · You may need different heating mantles for different sizes of flask

 ## THINGS TO DO AND THINK ABOUT

Sulfur dioxide gas is very soluble and dissolves to form an acidic solution. What would be the most appropriate method for collecting and measuring the volume of sulfur dioxide produced in a reaction?

 VIDEO LINK

Head to www.brightredbooks.net for a great video about the safe methods of heating listed in the table.

 DON'T FORGET

Alcohols and esters are highly flammable. They must never be heated using a Bunsen burner.

TECHNIQUES 2

FILTRATION

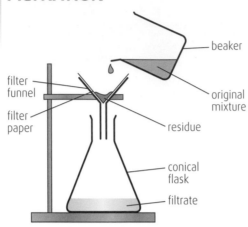

- beaker
- filter funnel
- original mixture
- filter paper
- residue
- conical flask
- filtrate

Filtration is used to separate a solid from a liquid.

DISTILLATION

Distillation is used to separate a mixture of liquids and can be used if the different liquids have different boiling points.

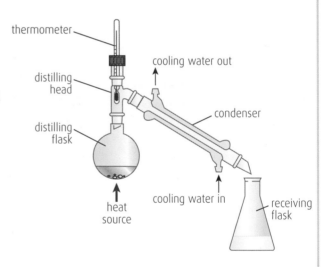

- thermometer
- cooling water out
- distilling head
- condenser
- distilling flask
- cooling water in
- heat source
- receiving flask

The mixture is gradually heated until it starts to boil. At this point, the compound with the lowest boiling point will turn into a vapour and pass out of the distilling flask into the condenser. When the vapour touches the cold walls of the condenser it turns back into a liquid and runs down to collect in the receiving flask.

If the mixture contains several different compounds, the temperature of the distilling flask can be increased gradually, allowing each compound to be collected in turn.

TITRATION

In a titration experiment, the concentration of a solution is measured by finding out the volume of the solution required to react with an accurately known amount of another chemical.

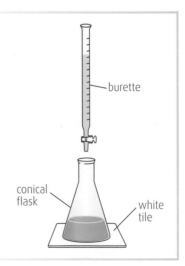

- burette
- conical flask
- white tile

Example:

Bottom burette reading / cm³	13·3	24·6	12·8
Top burette reading / cm³	0·5	12·4	0·5
Titre / cm³	12·8	12·2	12·3

Average titre = $\frac{12\cdot2 \times 12\cdot3}{2}$ = 12·25 cm³

Titrations are covered in detail on page 78.

USE OF A BALANCE

Chemical balances can be used to measure the mass of a sample accurately. The balance should be placed on a level surface well away from any vibrations or draughts. There are two different ways of using the balance.

Weighing chemicals directly into a beaker or flask

When using the balance to weigh out a certain mass of chemical for use in an experiment, the 'tare' button can be used. Place an empty beaker on the balance and press 'tare'. The balance will now adjust so that it reads zero. As you add the chemical to the beaker or flask, the balance will accurately record the mass of solid in the container.

Weighing by difference

Chemists will often weigh out solids into plastic or glass weighing boats or weighing bottles. When the solid is added to the reaction mixture, there is always a danger that traces of the solid will remain behind in the weighing boat. When it is important to know accurately how much solid has actually been used, a method called 'weighing by difference' is used. Using the 'tare' button to zero the balance with the weighing boat, you first roughly measure the correct mass of solid into the weighing boat. You then remove the weighing boat from the balance and press the 'zero' button. Once the balance has been properly zeroed, the weighing boat with the sample is placed on the balance and the total mass is recorded. The solid is then emptied from the weighing boat into the reaction vessel. The weighing boat with any traces of solid left behind is then weighed again. The actual mass of solid added is given by:

actual mass of solid used = (initial mass of boat + sample) − (final mass of boat + any remaining solid)

VIDEO LINK

For a video explaining how to weigh by difference, head to www.brightredbooks.net

EXPERIMENTAL UNCERTAINTIES

In chemical analysis, experiments are carried out to measure the concentration or mass of a particular element or compound present in a sample. One factor that can limit the accuracy of an analytical method is the choice of apparatus used.

Manufacturers will often label laboratory equipment to show how accurate the measurements made with the apparatus will be. The manufacturer of the pipette shown in the picture on the right guarantees that this pipette will measure out 25·00 ± 0·04 ml of liquid. This means that, if used correctly, the pipette will never measure out less than 25·00 − 0·04 = 24·96 ml and will never measure out more than 25·00 + 0·04 = 25·04 ml (1 ml is equal to 1 cm³).

The table shows three different pieces of apparatus that can be used to measure out 10 cm³ of a liquid.

Apparatus	Uncertainty
10 cm³ pipette	±0·04 cm³
10 cm³ measuring cylinder	±0·2 cm³
10 cm³ marking on a 25 cm³ beaker	±2 cm³

DON'T FORGET

The uncertainty in a measurement is an indication of how accurate the value is likely to be. It is shown by the ± sign followed by a number.

RELIABILITY

If an experiment is only carried out once, it is not possible to be sure whether the result obtained is completely reliable. For this reason, if time allows, chemists will repeat an experiment several times and record any measurements made. If all the values recorded are very close together, then the experiment is said to be highly reproducible. If, however, the values are spread over a fairly large range, then the experiment is said to have poor reproducibility.

DON'T FORGET

When a measurement has been repeated several times, look to see how close together the values are. This gives you an indication of the reliability of the measurement.

 THINGS TO DO AND THINK ABOUT

Beakers, measuring cylinders, pipettes and burettes can all be used to collect and measure liquids. What would be the most appropriate apparatus to collect: (a) roughly 50 cm³ of water to rinse a cloth sample; (b) a 20·00 cm³ sample of river water for use in a titration experiment; and (c) 22·3 cm³ of a standard acid solution needed to neutralise a sample.

ANALYSING RESULTS

REPRESENTING DATA USING A SCATTER GRAPH

Plotting graphs allows us to spot patterns in even fairly complicated experimental data.

Cooking temperature/°C	Time taken to tenderise beef/hours
71	36
75	24
79	15
85	9
88	8
92	5
96	3
98	2

For example, to investigate the changes in protein structure as beef is cooked, a series of experiments was carried out to find out how long it took samples of steak to become tender when cooked at different temperatures.

From this table, we can see that as the temperature is increased, the time taken for the meat to become tender decreases. If we plot a scatter graph of these data, we can see the pattern even more clearly.

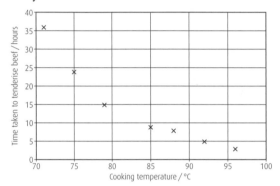

From the graph we can see that, at lower temperatures, a small increase in temperature results in a large decrease in cooking time. As we move to higher temperatures, to the right of the graph, each increase in temperature results in a smaller decrease in cooking time.

To show the trend seen in these data, we can add either a straight line of best fit, or a curve of best fit.

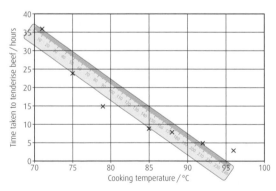

A straight line of best fit is not appropriate for these data.

DON'T FORGET

When drawing a line or curve of best fit, it does not have to pass through every point.

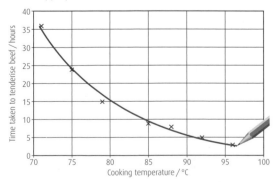

A smooth curve is drawn to show the trend in the results.

READING VALUES FROM A GRAPH OF EXPERIMENTAL DATA

The concentration of copper ions in a solution can be worked out by measuring how much light is absorbed by a sample. A set of five standard solutions containing different concentrations of copper ions was prepared and the amount of light they absorbed was measured. The data obtained were plotted as a scatter graph. With these data, a straight line of best fit is appropriate.

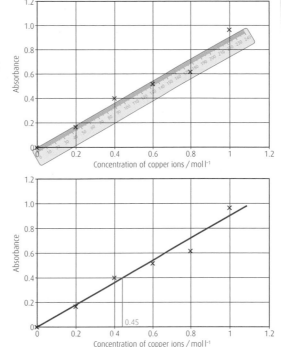

The line of best fit can now be used to work out the concentration of copper ions in other solutions.

If this sample has an absorbance of 0·40 units, the concentration of the solution can be read from the line of best fit. This solution would have a concentration of 0·45 mol l⁻¹.

DON'T FORGET

Use the line or curve of best fit when reading values from a graph.

ROGUE POINTS

Small errors in measurements are to be expected in experimental results. When data are plotted on a graph, the points will seldom lie on a perfectly straight line or on a perfectly smooth curve. When a measurement is repeated several times, small errors in the apparatus or measurements will, once again, lead to a slight variation in the data obtained.

The table on the right shows the concentration of calcium in samples of milk taken from a single bottle.

These five values are all reasonably close together and lie within a narrow range from 116 to 120 ppm. We can use these values to calculate the average concentration of calcium in milk.

Average calcium concentration = $\frac{117+119+120+116+118}{5}$ = 118 ppm

If, when you are looking at experimental data, there is a value that lies far away from all the other values, then it may be a rogue point. If time allows, you should repeat the experiment to see if the value obtained was reliable, but, if there is not sufficient time, this value should be removed from the calculation of averages.

Sulfite ions are added to wines as preservatives. The table on the right shows the concentrations of sulfite ion in samples taken from a single bottle of white wine.

Here, all but one of the values lie in a reasonably narrow range between 19·4 and 22·1 ppm. The result for sample 4 is almost four times larger than the other values. Because this value is not typical of the wine being tested, it is not included in the average value.

Average sulfite concentration = $\frac{19·4+22·1+20·4+21·6}{4}$ = 20·9 ppm

DON'T FORGET

Do not include rogue points when calculating averages.

Sample	Concentration of calcium/ppm
1	117
2	119
3	120
4	116
5	118

Sample	Concentration of sulfite ions/ppm
1	19·4
2	22·1
3	20·4
4	84·2
5	21·6

 THINGS TO DO AND THINK ABOUT

For each of the following experiments, calculate the average sodium content.

a   Five samples were taken from a single block of butter and analysed. The samples contained 48·9 ppm, 49·2 ppm, 52·1 ppm, 48·1 ppm and 51·9 ppm of sodium.

b   Five samples of seawater were taken from a single location at a single time. The samples contained 1030 ppm, 1015 ppm, 1040 ppm, 1990 ppm and 1025 ppm of sodium.

 VIDEO LINK

Learn more about how to plot a graph and add a line or curve to describe the trend seen at www.brightredbooks.net

ANSWERS

CONTROLLING THE RATE OF A CHEMICAL REACTION 1

1 Methane and oxygen gas do not react together when they are cold because their molecules are moving too slowly and do not collide with sufficient energy. A hot match or lighter increases the temperature of the gas mixture, making the molecules speed up. The faster moving molecules now collide with enough energy to react and the chemical reaction starts. Methane burning is an exothermic reaction and the heat given out provides the remaining particles with enough energy to collide successfully.

2 If the heat energy released in an exothermic reaction is not removed, the temperature of the reaction mixture will rise. The more the temperatures rises, the faster the chemical reaction will proceed and the faster the heat energy will be released. Eventually a point may be reached at which the heat energy is being released so quickly that an explosion occurs.

CONTROLLING THE RATE OF A CHEMICAL REACTION 2

For a reaction to occur, particles must collide with sufficient kinetic energy. (They must collide with an energy equal to, or greater than, the activation energy.)

The particles must also collide with the correct orientation or alignment.

CONTROLLING THE RATE OF A CHEMICAL REACTION 3

A catalyst lowers the activation energy of the reaction and so more reacting particles have an energy greater than, or equal to, the lowered activation energy. Increasing the temperature gives the reacting particles more energy and so more reacting particles have an energy equal to, or greater than, the activation energy. This time the activation energy has not been lowered.

PERIODICITY 1

Using the graph showing the boiling points of the noble gases on page 13, we can see that the points lie on a roughly straight line. Using a line of best fit (see page 87), we could predict that a noble gas with 118 electrons would have a boiling point of around +35°C.

PERIODICITY 2

The intermolecular forces or bonds being broken are as follows: sulfur, London dispersion forces; zinc, metallic bonds; phosphorus, London dispersion forces; argon, London dispersion forces; chlorine, London dispersion forces; carbon in the form of diamond, covalent bonds; and carbon in the form of fullerene, London dispersion forces.

PERIODICITY 3

(i) Lithium atoms only have three electrons. The fourth ionisation energy would require the removal of a fourth electron from a lithium atom.

(ii) Aluminium atoms have the electronic arrangement 2, 8, 3. The first three ionisation energies measure the energy required to remove electrons from the third shell. The fourth ionisation energy measures the energy required to remove an electron from the second shell. Electrons in the second shell are closer to the nucleus and are less shielded from the nuclear charge as there are fewer underlying electron shells.

STRUCTURE AND BONDING 1

1   Using the differences in electronegativity values, you would predict: NaI, difference 1·7, polar covalent; CaF_2, difference 3·0, ionic; SbI_3, difference 0·5, polar covalent; and $AuBr_3$, difference 0·4, polar covalent.

2   Antimony bromide is a covalent molecular substance containing polar covalent bonds.

STRUCTURE AND BONDING 2

1   Ammonia, hydrogen cyanide and trichloromethane are all polar.

2   The following three molecules will have hydrogen bonds between their molecules:

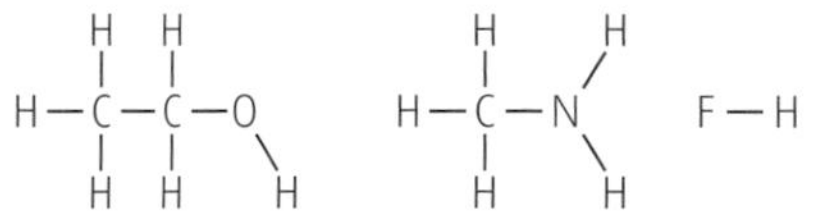

STRUCTURE AND BONDING 3

The molecules listed in order from the least viscous to the most viscous are: butane; 2-methoxypropane; butan-2-ol; and butane-2,3-diol.

ANSWERS TO UNIT 2: NATURE'S CHEMISTRY

ESTERS, FATS AND OILS 1

1   a   ethyl butanoate   b   butyl ethanoate

2   a   propyl ethanoate   b   ethyl pentanoate
     c   ethyl butanaote

3   a 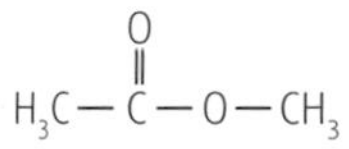
methyl ethanoate

b 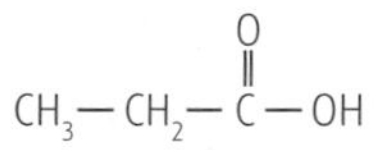
propanoic acid

ESTERS, FATS AND OILS 3

1   octadec-9,12,15-trienoic acid

PROTEINS

1   B

CHEMISTRY OF COOKING 1

1   A

2   A

OXIDATION OF FOOD 1

1   B

OXIDATION OF FOOD 2

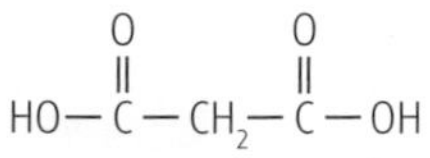

ANSWERS TO UNIT 2: NATURE'S CHEMISTRY

FRAGRANCES

1

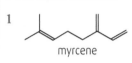

myrcene

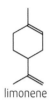

limonene

SKIN CARE

1 a Initiation.

   b $CH_3\bullet + F_2 \rightarrow CH_3F + F\bullet$

   c $CH_3\bullet + CH_3\bullet \rightarrow CH_3CH_3$ and $F\bullet + F\bullet \rightarrow F_2$

2 Compound **A** can be described as a free radical scavenger because it reacts with a free radical to form a stable molecule.

ANSWERS TO UNIT 3: CHEMISTRY IN SOCIETY

GETTING THE MOST FROM REACTANTS 2

1 a Because oxygen is in excess, all the magnesium will be converted into magnesium oxide.

   The number of moles of Mg used = $\frac{mass}{GFM} = \frac{4\cdot86}{24\cdot3} = 0\cdot2$ mol

   The balanced chemical equation for the reaction is:

   $2Mg + O_2 \rightarrow 2MgO$

   From this we can see that 2 mol of Mg will produce 2 mol of MgO.

   So if 0·2 mol of Mg is used up, then 0·2 mol of MgO will be produced.

   Mass of MgO = $n \times GFM = 0\cdot2 \times 40\cdot3 = $ **8·06 g**

   b The number of moles of $CaCO_3$ heated = $\frac{mass}{GFM} = \frac{2\cdot05}{100\cdot1}$ = 0·0205 mol

   The balanced chemical equation for the reaction is:

   $CaCO_3 \rightarrow CaO + CO_2$

   From this we can see that 1 mol of $CaCO_3$ will produce 1 mol of CaO.

   So if 0·0205 mol of $CaCO_3$ are used up, then 0·0205 mol of CaO will be produced.

   Mass of CaO = $n \times GFM = 0\cdot0205 \times 56\cdot1 = $ **1·15 g**

2 Because they are all gases, the value which has the greater number of moles of gas will occupy the largest volume.

   Number of moles, n = mass/GFM

   Remember hydrogen, oxygen and nitrogen are diatomic gases.

   a $n = 0\cdot2/2 = 0\cdot1$ mol; b $n = 3\cdot2/32 = 0\cdot1$ mol; c $n = 0\cdot8/4\cdot0 = 0\cdot2$ mol; and d $n = 2\cdot8/28 = 0\cdot1$ mol. Therefore (c), 0·8 g of He, occupies the greatest volume.

3 One mole of water has a mass of 18·0 g and contains $6\cdot02 \times 10^{23}$ water molecules.

   Therefore 180 g of water will contain $6\cdot02 \times 10^{23} \times 180/18\cdot0 = $ **$6\cdot02 \times 10^{24}$** H_2O molecules.

GETTING THE MOST FROM REACTANTS 3

1 The balanced chemical equation for ethene burning completely is:

   $C_2H_4(g) + 3O_2(g) \rightarrow 2CO_2(g) + 2H_2O(l)$

   1 vol   + 3 vol   → **2 vol**   (ignore water because it is a liquid)

   200 cm³ + 600 cm³ → 400 cm³

   a The volume of excess oxygen = 900 − 600 = 300 cm³

   b The total volume of gas at the end = 400 cm³ CO_2 + 300 cm³ unreacted O_2 = **700 cm³**

2 The number of moles of $MgCO_3$ = $\frac{mass}{GFM} = \frac{0\cdot281}{84\cdot3}$ = 0·00333 mol.

   From the balanced chemical equation, you can see that

contd

1 mol of $MgCO_3$ will produce 1 mol of CO_2.

So the number of moles of CO_2 is also 0·00333 mol.

Therefore the volume produced = 0·00333 × 22·4 = **0·0746 l or 74·6 cm³**.

3   The number of moles of Zn = $\frac{mass}{GFM}$ = $\frac{1·04}{65·4}$ = 0·0159 mol.

From the balanced chemical equation, you can see that 1 mol of Zn will produce 1 mol of H_2.

So the number of moles of H_2 is also 0·0159 mol.

Therefore the volume produced = 0·0159 × 22·4 = **0·356 l or 356 cm³**.

GETTING THE MOST FROM REACTANTS 5

2   The aluminium chloride catalyst in the Boots' synthesis is not a true catalyst and is disposed of in landfill sites. This means that fresh aluminium chloride has to be used for the next batch. The catalysts used in the BHC synthesis are all true catalysts and can be recovered and reused.

3   In the Haber Process, any unreacted nitrogen and hydrogen are continuously recirculated through the reaction chamber over the iron catalyst, thus increasing the percentage yield.

GETTING THE MOST FROM REACTANTS 6

1   Number of moles of sulfuric acid, $n = cV$ = 0·20 × 0·1 = 0·020 mol

The balanced equation is:
$CuCO_3 + H_2SO_4 \rightarrow CuSO_4 + H_2O + CO_2$

Therefore 1 mol of sulfuric acid reacts with 1 mol of $CuCO_3$ and 0·020 mol of copper(II) carbonate will be required to react completely with 0·020 mol of $CuCO_3$.

GFM of $CuCO_3$ = 123·5 g, so 1 mol = 123·5 g, therefore 0·020 mol = 123·5 × 0·020 = **2·47 g**

2   a n for MgO = $\frac{mass}{GFM}$ = $\frac{2·015}{40·3}$ = 0·05 mol; n for HCl = cV = 0·101 × 0·1 = 0·0101 mol

The balanced equation is
$MgO + 2HCl \rightarrow MgCl_2 + H_2O$ and so 0·05 mol of Mg would react with 2 × 0·05 = 0·10 mol of HCl.

There is only 0·0101 mol of HCl present, so MgO is in excess and HCl is the limiting reactant.

From the above equation we can see that 2 mol of HCl produces 1 mol of the salt $MgCl_2$.

Therefore 0·0101 mol of HCl will produce $\frac{0·0101}{2}$ = 0·00505 mol of $MgCl_2$.

Mass of $MgCl_2$ formed = n × GFM = 0·00505 × 95·3 = **0·48 g**

b n for $MgCO_3$ = 4·000/84·3 = 0·0474 mol; n for HNO_3 = 0·010 × 0·2 = 0·002 mol

The balanced equation is
$MgCO_3 + 2HNO_3 \rightarrow Mg(NO_3)_2 + H_2O + CO_2$

$MgCO_3$ is in excess and the limiting reagent is HNO_3.

0·002 mol of HNO_3 will produce 0·001 mol of the salt $Mg(NO_3)_2$.

Mass of $Mg(NO_3)_2$ = n × GFM = 0·001 × 148·3 = **0·1483 g**

c n for Zn = $\frac{2·616}{65·4}$ = 0·04 mol; n for H_2SO_4 = 0·210 × 0·25 = 0·0525 mol

The balanced equation is $Zn + H_2SO_4 \rightarrow ZnSO_4 + H_2$

H_2SO_4 is in excess and the limiting reagent is Zn.

0·04 mol of Zn will produce 0·04 mol of the salt $ZnSO_4$.

Mass of $ZnSO_4$ = n × GFM = 0·04 × 161·5 = **6·46 g**

CHEMICAL ENERGY 1

1   The enthalpy of combustion is the heat given out when 1 mole is burned completely in oxygen and the units are kJ mol⁻¹.

The GFM of methanol is 32 g.

0·14 g of methanol gives out 3·04 kJ

Therefore 32 g gives out (3·04 × 32)/0·14 = 695 kJ

Therefore ΔH_c = **–695 kJ mol⁻¹** (remember to put in the negative sign because it is an exothermic reaction).

2   Mass of water = 164·3 – 62·3 = 102 g = 0·102 kg

ΔT = 28·5 – 18·0 = 10·5°C

$E_h = cm\Delta T$ = 4·18 × 0·102 × 10·5 = 4·48 kJ

Mass of propan-1-ol combusted = 24·36 – 24·14 g = 0·22 g

GFM of propan-1-ol ($CH_3CH_2CH_2OH$) = 60 g

So 0·22 g burns to give out 4·48 kJ

60 g will burn to give out 4·48 × 60/0·22 = 1222 kJ

Therefore ΔH_c of propan-1-ol = **–1222 kJ mol⁻¹**

The two main reasons why this value is much lower than that given in the Data Booklet are:

- much of the heat given out by the burning propan-1-ol will be lost to the surrounding air;

- the propan-1-ol will have been burned in air rather than in pure oxygen and so there is likely to have been incomplete combustion of the propan-1-ol.

contd

ANSWERS TO UNIT 3: CHEMISTRY IN SOCIETY

3 • There will be virtually no heat lost to the surrounding air.

• The alcohol will be burning in oxygen rather than

in air, which is only 21% oxygen, and so complete combustion is more likely to take place.

• As it is a closed system, there is no possibility of the propan-1-ol evaporating rather than burning.

CHEMICAL ENERGY 3

1 The relevant equations and their enthalpy values from the Data Booklet are:

1 $C_2H_4 + 3O_2 \rightarrow 2CO_2 + 2H_2O$ $\Delta H = -1411$ kJ
2 $H_2 + \frac{1}{2}O_2 \rightarrow H_2O$ $\Delta H = -286$ kJ
3 $C_2H_6 + 3\frac{1}{2}O_2 \rightarrow 2CO_2 + 3H_2O$ $\Delta H = -1561$ kJ

To obtain the desired equation, Equation ③ has to be reversed and added to Equations ① and ②.

Therefore ΔH for the desired equation =
$-1411 - 286 + 1561 = -136$ kJ mol^{-1}.

2 The equation showing the bonds is:

$$H_2C=CH_2 + H-H \longrightarrow H-CH_2-CH_2-H$$

Bond-breaking (ΔH +ve)

$1 \times$ C=C, ΔH = + 612 kJ

$1 \times$ H–H, ΔH = +436 kJ

Bond-making (ΔH –ve)

$1 \times$ C–C, ΔH = –348 kJ

$2 \times$ C–H, $\Delta H = 2 \times (-412) = -824$ kJ

So the overall ΔH = 612 + 436 – 348 – 824 = –124 kJ mol^{-1}.

3 This is because the bond enthalpy values are mean values and, for example, the bond enthalpy values for the C–C and C–H bonds in ethane may be slightly different from the mean values given in the Data Booklet.

4 The equation showing the bonds is:

$$H-H + \tfrac{1}{2}(O=O) \rightarrow \; H\!-\!O\!-\!H$$

Bond-breaking (ΔH +ve)

$1 \times$ H–H, ΔH = +436 kJ

$\frac{1}{2} \times$ O=O, $\Delta H = \frac{1}{2} \times 498 = +249$ kJ

Bond-making (ΔH –ve)

$2 \times$ O–H $= 2 \times (-463) = -926$ kJ

So the overall ΔH = 436 + 249 – 926 = –241 kJ mol^{-1}.

Reasons for the difference are:

• Bond enthalpy values from the Data Booklet are mean values and the O–H bond in water may have a different value from, say, the O–H bond in alcohols.

• The equation for the enthalpy of combustion of hydrogen gives liquid water as the product, whereas the equation you used in this calculation has water in the gas state as the product.

5 a The equation showing the bonds is

$$4\,(N{\equiv}N) \rightarrow$$

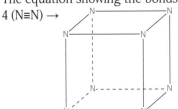

Bond-breaking (ΔH +ve)

$4 \times$ N≡N $= 4 \times 945 = +3780$ kJ

Bond-making (ΔH –ve)

$12 \times (-163) = -1956$ kJ

So ΔH for the reaction
= 3780 – 1956 = +1824 kJ mol^{-1}.

b The bond angles are 90° which would make N_8 unstable and this is confirmed by the large positive enthalpy value.

OXIDISING AND REDUCING AGENTS 1

1 a $Cl_2(g) + 2e^- \rightarrow 2Cl^-(aq)$

b $MnO_4^-(aq) + 8H^+(aq) + 5e^- \rightarrow Mn^{2+}(aq) + 4H_2O(l)$

c $Zn(s) \rightarrow Zn^{2+}(aq) + 2e^-$

d $SO_3^{2-}(aq) + H_2O(l) \rightarrow SO_4^{2-}(aq) + 2H^+(aq) + 2e^-$

e $Cr_2O_7^{2-}(aq) + 14H^+(aq) + 6e^- \rightarrow 2Cr^{3+}(aq) + 7H_2O(l)$

OXIDISING AND REDUCING AGENTS 2

a Bromine will oxidise the iodide ions to iodine. The ion-electron equations are:

$2I^- \rightarrow I_2 + 2e^-$ Oxidation

$Br_2 + 2e^- \rightarrow 2Br^-$ Reduction

$Br_2 + 2I^- \rightarrow 2Br^- + I_2$ Redox

b No reaction will take place.

c Zinc will displace copper. A displacement reaction is a redox reaction.

$Zn \rightarrow Zn^{2+} + 2e^-$ Oxidation

$Cu^{2+} + 2e^- \rightarrow Cu$ Reduction

$Zn + Cu^{2+} \rightarrow Zn^{2+} + Cu$ Redox

d No reaction will take place.

e The dichromate ions and the sulfite ions will react.

$(SO_3^{2-} + H_2O \rightarrow SO_4^{2-} + 2H^+ + 2e^-) \times 3$ Oxidation

$Cr_2O_7^{2-} + 14H^+ + 6e^- \rightarrow 2Cr^{3+} + 7H_2O$ Reduction

$Cr_2O_7^{2-} + 8H^+ + 3SO_3^{2-} \rightarrow 2Cr^{3+} + 3SO_4^{2-} + 4H_2O$ Redox

CHEMICAL ANALYSIS 1

A Because there is only one spot corresponding to the sample and so it must be a pure compound.

ANSWERS TO UNIT 4: RESEARCHING CHEMISTRY

LABORATORY APPARATUS

Volumetric flasks are used to prepare solutions of accurately known concentrations. It is very important that the only chemicals they contain are those carefully measured into the flask when the solution is being prepared. When the flask is rinsed with distilled water before use it does not matter if a few drops of distilled water are left behind because distilled water is going to be added later anyway, as the total volume of the solution is made up to the line.

TECHNIQUES 1

Because sulfur dioxide is soluble, it cannot be collected over water. A gas syringe must be used to collect and measure the volume of this gas.

TECHNIQUES 2

a When collecting water to rinse a cloth sample, the actual volume used is not important so the rough markings on a beaker would be appropriate for this experiment.

b In titrations, the concentration of substances are measured accurately. A 20 cm³ pipette would give the most accurate measurement. Another clue that a pipette should be used is that the volume of the solution, in cm³, is quoted to two decimal places.

c Pipettes are only available in certain standard sizes, so a burette would have to be used to measure 22·3 cm³. Another clue that a burette is being used is that the volume of the solution, in cm³, is quoted to one decimal place.

ANALYSING RESULTS

a All five values are spread across a fairly narrow range (between 48·9 and 52·1 ppm) and there are no obvious rogue points. All five values are used to calculate an average of 50·0 ppm.

b Four of the values are spread across a fairly narrow range (between 1015 and 1040 ppm), but the remaining value is almost twice as large as the others. The value 1990 ppm is therefore a rogue point and is not included in the average. The average calculated from the four remaining values is 1028 ppm.

INDEX